AutoCAD® Pocket Reference
4th Edition

by
Cheryl R. Shrock
Professor
Drafting Technology
Orange Coast College, Costa Mesa, Ca.
and
Autodesk® Authorized Author

INDUSTRIAL PRESS
New York

Industrial Press Inc.
989 Avenue of the Americas
New York, NY 10018

10 9 8 7 6 5 4 3 2

Why do you need this book?

Refresh your memory or learn something new.
No need to memorize.
Handy size and easy to use.

About this book

The *AutoCAD Pocket Reference, 4th Edition*, includes all the important fundamental Commands, Concepts, and How to information for the every day use of AutoCAD. It is not designed to take the place of larger textbooks but rather to supplement them as a quick reference.
Note: If you have an earlier version of AutoCAD software refer to the AutoCAD Pocket Reference published previously to this edition.

How to use this book

The information in this book has been organized in 12 sections. Each section contains related material. For example, if you needed information regarding dimensioning, you would go to:
Section 3: Dimensioning.
A comprehensive Table of Contents and a cross-referenced Index have been carefully prepared to insure easy access to all information.

Words from the Author

"I originally wrote this book for myself. Occasionally I forget commands and "How to" steps also. Its convenient small size allows me to toss it in my briefcase or set it beside my monitor. No need to memorize anymore."

Cheryl R. Shrock is a Professor and Chairperson of Computer Aided Design at Orange Coast College in Costa Mesa, California.
She is also an Autodesk® registered
author / publisher and has written 12 versions of
"Exercise Workbook for AutoCAD".
Cheryl is always trying to think of new ways to make it easy to learn AutoCAD. This Pocket Reference is the latest in that endeavor.

AutoCAD Books by Cheryl R. Shrock:

Beginning AutoCAD **2006**	ISBN 978-08311-3213-2
Advanced AutoCAD **2006**	ISBN 978-08311-3214-9
Beginning AutoCAD **2007**	ISBN 978-08311-3302-3
Advanced AutoCAD **2007**	ISBN 978-08311-3303-0
Beginning AutoCAD **2008**	ISBN 978-0-8311-3341-2
Advanced AutoCAD **2008**	ISBN 978-0-8311-3342-9
Beginning AutoCAD **2009**	ISBN 978-0-8311-3359-7
Advanced AutoCAD **2009**	ISBN 978-0-8311-3360-3
Beginning AutoCAD **2010**	ISBN 978-0-8311-3404-4
Advanced AutoCAD **2010**	ISBN 978-0-8311-3400-6
AutoCAD Pocket Reference **2006, Releases 2006- 2004**	ISBN 978-0-8311-3263-7
AutoCAD Pocket Reference **2007, Releases 2007- 2004**	ISBN 978-0-8311-3328-3
AutoCAD Pocket Reference **2008, Releases 2008-2006**	ISBN 978-0-8311-3354-2
AutoCAD Pocket Reference **2009, Releases 2009 or later**	ISBN 978-0-8311-3384-9

For information about these books visit www.industrialpress.com

For information about Cheryl Shrock's online courses, visit www.shrockpublishing.com

TABLE OF CONTENTS

The following Table of Contents is an overview of the contents within this book. Refer to the Index at the back of this book to easily locate specific information.

SECTION 1

Action Commands

SECTION 2

Concepts

SECTION 3

Dimensioning

SECTION 4

Drawing Entities

SECTION 5

How to...

SECTION 6

Layers

SECTION 7

Input options

SECTION 8

Miscellaneous

SECTION 9

Plotting

SECTION 10

Settings

SECTION 11

Text

SECTIONS 12

UCS

SECTIONS 13

CONSTRAINTS

Section 1
Action Commands

ARRAY

The ARRAY command allows you to make multiple copies in a **RECTANGULAR** or
Circular **(POLAR)** pattern. The maximum limit of copies per array is 100,000. This
limit can be changed but should accommodate most users. (Refer to Help menu)

RECTANGULAR ARRAY
This method allows you to make multiple copies of object(s) in a rectangular pattern.
You specify the number of rows (horizontal), columns (vertical) and the offset distance
between the rows and columns. The offset distances will be equally spaced.

Offset Distance is sometimes tricky to understand. ***Read this carefully***. The offset
distance is the <u>distance from a specific location on the original to that same location</u> on
the future copy. It is <u>not just the space in between </u>the two. Refer to the example
below.

To use the rectangular array command you will select the object(s), specify how many
rows and columns desired and the offset distance for the rows and the columns.
<u>**Refer to step by step instructions on page 1- 3.**</u>

Example of Rectangular Array:

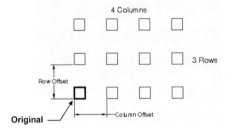

Example of Rectangular Array on an angle:
Notice the copies do not rotate. The Angle is only used to establish the placement.

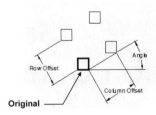

ARRAY….continued

RECTANGULAR ARRAY (Step by step)

1. Select the **ARRAY** command using one of the following:

    Ribbon = Home tab / Modify panel / 🔲

    Keyboard = Array <enter>

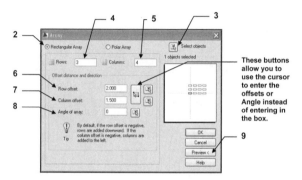

2. Select "**Rectangular Array**".

3. Select the **"Select Objects"** button.
   This will take you back to your drawing. Select the objects to Array then <enter>.

4. Enter the number of rows. (Horizontal rows)

5. Enter the number of columns. (Vertical columns)

6. Enter the row offset. (<u>The distance from a specific location on the original to that same specific location on the future copy.</u>) See example on pg. 1-2.

7. Enter the column offset. (<u>The distance from a specific location on the original to that same specific location on the future copy.</u>) See example on pg. 1-2.

8. Enter an angle if you would like the array to be on an angle.

9. Select the **Preview** button.
   <u>If it looks correct</u>, right click. <u>If it is **not** correct</u>, press the **ESC** key, make the necessary corrections and preview again.
   (Note: If the Preview button is gray, you have forgotten to "Select objects"- #3)

ARRAY....continued

POLAR ARRAY

This method allows you to make multiple copies in a <u>circular pattern</u>. You specify the total number of copies to fill a specific Angle or specify the angle between each copy and angle to fill.

To use the polar array command you select the object(s) to copy, specify the center of the array, specify the number of copies or the angle between the copies, the angle to fill and if you would like the copies to rotate as they are copied.

Example of Polar Array

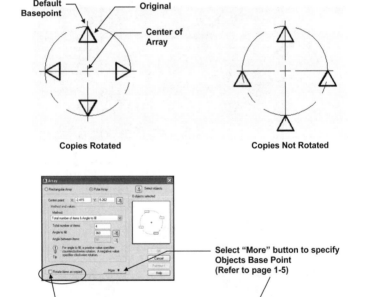

Default Basepoint

Original

Center of Array

Copies Rotated

Copies Not Rotated

Rotate Option

Select "More" button to specify Objects Base Point
(Refer to page 1-5)

ARRAY....continued

Note: the two examples shown on the previous page use the **objects default base point**. The examples below displays what happens if you specify a basepoint.

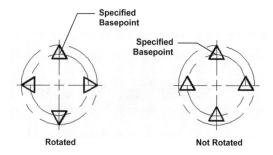

Rotated   Not Rotated

List of Default Base Point locations
Arc, Circle and Ellipse = Center of Object
Polygon and Rectangle = First Corner
Line, Polyline and Donut = Starting Point
Text and Block = Insertion Point

Note: If you select multiple objects the base point of the last object selected is used to construct the Array.

Uncheck box →

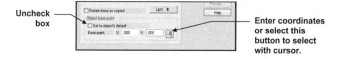

← Enter coordinates or select this button to select with cursor.

The difference between "Center Point" and "Object Base Point" is sometimes confusing.

Center Point
When specifying the Center Point, try to visualize the copies already there. Now, in your mind, try to visualize the center of that array. In other words, it is the Pivot Point from which the copies will be placed around.

Base Point
The Base Point is different. It is located on the original object. You must select the Base Point in addition to the Center Point. They are two different options.

ARRAY....continued

POLAR ARRAY (Step by step)

1. Select the ARRAY command using one of the following:

   Ribbon = Home tab / Modify panel /

   Keyboard = Array <enter>

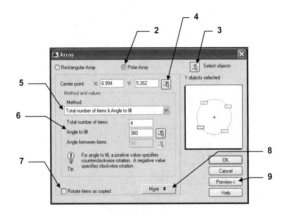

2. Select **" Polar Array"**.

3. Select the **"Select Objects"** button.
   This will take you back to your drawing. Select the objects to Array then <enter>.

4. Select the "Center Point" button and select the center point with the cursor or enter the X and Y coordinates in the "Center Point" boxes.

5. Select the method. (Examples on the next page)

6. Enter the "Total number of items", "Angle to fill" or "Angle between items".

7. Select whether you want the items rotated as copied or not.

8. Accept the object's default base point or enter the X and Y coordinates.
   (Select the "More" button to show this area. Select the "Less" button to not show.)

9. Select the Preview button.
   If it looks correct, right click. If it is **not** correct, press the ESC key, make the necessary corrections and preview again.
   (Note: If the Preview button is gray, you have forgotten to "Select objects"- #3)

ARRAY....continued

Total number of items & <u>Angle to fill</u>

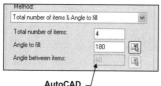

AutoCAD
Calculates
<u>Angle between items</u>

Rotated   Not Rotated

Total number of items & <u>Angle between items</u>

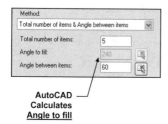

AutoCAD
Calculates
<u>Angle to fill</u>

Rotated   Not Rotated

<u>Angle to fill</u> & <u>Angle between items</u>

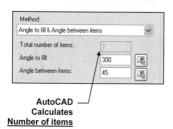

AutoCAD
Calculates
<u>Number of items</u>

Rotated   Not Rotated

BREAK

The **BREAK** command allows you to break an object at a single point (Break at Point) or between two points (Break). I think of it as breaking a single line segment into two segments or taking a bite out of an object.

METHOD 1 - Break at a Single Point
How to break one Line into two separate objects with no visible space in between.

1. Select the **BREAK AT POINT** command by using::

   Ribbon = Home tab / Modify panel / [image]

2. _break Select objects: *select the object to break (P1).*

3. Specify first break point: *select break location (P2) accurately.*

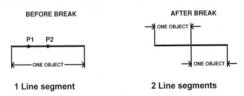

BEFORE BREAK

AFTER BREAK

1 Line segment

2 Line segments

BREAK....continued

METHOD 2
Break between 2 points. (Take a bite out of an object.)
(Use this method if the <u>location of</u> the BREAK <u>is not important</u>.)

1. Select the **BREAK** command using one of the following:

    Ribbon = Home tab / Modify panel / 🔲

    Keyboard = BR <enter>

2. _break Select object: *pick the first break location (P1).*
3. Specify second break point or [First point]: *pick the second break location (P2).*

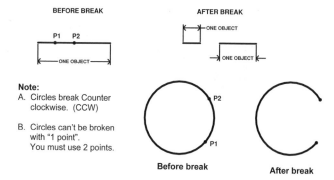

Note:
A. Circles break Counter
    clockwise. (CCW)

B. Circles can't be broken
    with "1 point".
    You must use 2 points.

The following method is the same as method 2 above; however, <u>use this method
if the location of the break is very specific.</u>

1. Select the **BREAK** command.
2. _break Select objects: *select the object to break (P1) anywhere on the object.*
3. Specify second break point or [First point]: *type F <enter>.*
4. Specify first break point: *select the first break location (P2) accurately.*
5. Specify second break point: *select the second break location (P3) accurately.*

CHAMFER

The **CHAMFER** command allows you to create a chamfered corner on two lines.
There are two methods: **Distance (below) and Angle (next page)**.

DISTANCE METHOD

Distance Method requires input of a distance for each side of the corner.

1. Select the **CHAMFER** command using one of the following:

   Ribbon = Home tab / Modify panel /

   Keyboard = CHA <enter>

   Command: _chamfer
   (TRIM mode) Current chamfer Dist1 = 0.000, Dist2 = 0.000
   Select first line or [Undo/Polyline/Distance/Angle/Trim/mEthod/Multiple]: *select "D"<enter>.*
   Specify first chamfer distance <0.000>: *type the distance for first side <enter>.*
   Specify second chamfer distance <1.000>: *type the distance for second side <enter>.*

2. **Now chamfer the object.**
   Select first line or [Undo/Polyline/Distance/Angle/Trim/mEthod/Multiple]: *select the*
   (First side) to be chamfered (distance 1).
   Select second line or shift-select to apply corner: *select the (Second side) to be*
   chamfered (distance 2).

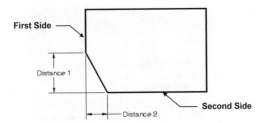

Polyline: This option allows you to Chamfer all intersections of a Polyline in one operation.
Such as all 4 corners of a rectangle.

Trim: This option controls whether the original lines are trimmed or remain after the corners
are chamfered. (Set to Trim or No trim.)

mEthod: Allows you to switch between **Distance** and **Angle** method. The distance or angle
must have been set previously.

Multiple: Repeats the Chamfer command until you press <enter> or esc key.

CHAMFER....continued

ANGLE METHOD

Angle method requires input for the length of the line and an angle

1. Select the **CHAMFER** command

   Command: _chamfer
   (TRIM mode) Current chamfer Dist1 = 1.000, Dist2 = 1.000
   Select first line or [Undo/Polyline/Distance/Angle/Trim/method/Multiple]: ***type A <enter>***
   Specify chamfer length on the first line <0.000>: ***type the chamfer length <enter>***
   Specify chamfer angle from the first line <0>: ***type the angle <enter>***

2. **Now Chamfer the object**
   Select first line or [Undo/Polyline/Distance/Angle/Trim/mEthod/Multiple]: ***select the (First Line) to be chamfered. (the length side)***
   Select second line or shift-select to apply corner: ***select the (second line) to be chamfered. (the Angle side)***

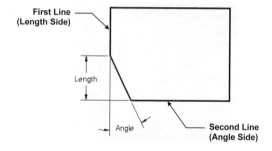

First Line (Length Side)

Length

Angle

Second Line (Angle Side)

COPY

The **COPY** command creates a duplicate set of the objects selected.
The COPY command is similar to the MOVE command.

The steps required are:
    1. Select the objects to be copied,
    2. Select a base point
    3. Select a New location for the New copy.

The difference between Copy and Move commands:
The Move command merely moves the objects to a new location.
The Copy command makes a copy and you select the location for the new copy.

1. Select the Copy command using one of the following commands:

    Ribbon = Home tab / Modify panel /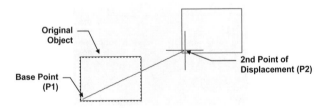

    Keyboard = CO <enter>

2. The following will appear on the command line:

Command: _copy
Select objects: *select the objects you want to copy*
Select objects: *stop selecting objects by selecting <enter>*
Current settings: Copy mode = Multiple
Specify base point or [Displacement/mOde] <Displacement>: *select a base point (P1)*
Specify second point of displacement or <use first point as displacement>: *select the new location (P2) for the first copy*
Specify second point or [Exit / Undo] <Exit>: *select the new location (P2) for the next copy or press <enter> to exit.*

Original
Object

Base Point
(P1)

2nd Point of
Displacement (P2)

Continued on the next page...

COPY....continued

Note:
If you select the option **mOde**, you may select <u>Single</u> or <u>Multiple</u> copy mode.
The default setting is Multiple.
It is practical to leave the mode setting at Multiple. If you choose to make only one copy
just press <enter> to exit the Copy command.

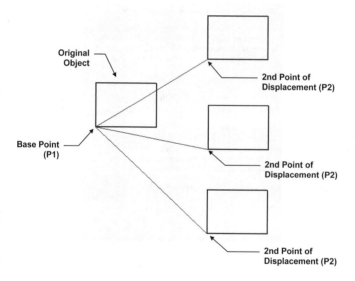

The copy command continues to make copies until you press <enter> to exit.

DIVIDE COMMAND

The DIVIDE command divides an object mathematically by the NUMBER of segments you specify. A POINT (object) is placed at each interval on the object.

Note: the object selected is **NOT** broken into segments. The **POINTS** are simply drawn **ON** the object.

EXAMPLE:

This LINE has been DIVIDED into 4 EQUAL lengths.
But remember, the line is not broken into segments.
The Points are simply drawn **ON** the object.

1. First open the Point Style box and select the **POINT STYLE** to be placed on the object.

    Keyboard = ddptype <enter>

        Select either of these ONLY when you want to make the points disappear yet still be there.

(Refer to the Point command if you need a refresher on Points)

2. Next select the **DIVIDE** command using one of the following:

    Ribbon = Home tab / Draw panel ▾ /  (on the Point tool)

    Keyboard = DIV <enter>

3. Select Object to divide: *select the object to divide*

4. Enter the number of segments or [Block]: *type the number of segments <enter>*

ERASE

There are 3 methods to erase (delete) objects from the drawing.
They all work equally well. You decide which one you prefer to use.

Method 1
Select the Erase command first and then select the objects.

Example:
1. Start the Erase command using one of the following:

Ribbon = Home tab / Modify panel /

Keyboard = E <enter>

2. Select objects: ***Pick one or more objects***
    Select Objects: ***Press <enter> and the objects selected will disappear.***

Method 2
Select the Objects first and then the Delete Key.

Example:
1. Select the object to be erased.
2. Press the Delete Key.

Method 3
Select the Objects first and then select Erase command from the Shortcut Menu.

Example:
1. Select the object to be erased.
2. Press the right Mouse button.
3. Select Erase from the Shortcut Menu.

Note: Very Important
If you want the erased objects to return, select the **Undo tool** from the **Quick Access Toolbar**. This will Undo the last command.
More about Undo and Redo later.

EXPLODE

The **EXPLODE** command changes (explodes) an object into its primitive objects.
For example: A rectangle is originally one object. If you explode it, it changes into 4 lines.
Visually you will not be able to see the change unless you select one of the lines.

1. Select the **Explode** command by using one of the following:

   Ribbon = Home tab / Modify panel /

   Keyboard = X <enter>

2. The following will appear on the command line:

Command: _explode
Select objects: *select the object(s) you want to explode.*
Select objects: *select <enter>.*

Before EXPLODE   **After EXPLODE**

One Object   **4 Objects**
(Rectangle)   **(4 Lines)**

**(Notice there is no visible difference. But now you have 4 lines instead of 1
Rectangle)**

Try this:
Draw a rectangle and then click on it. The entire object highlights.
Now explode the rectangle, then click on it again. Only the line you clicked on should be
highlighted. Each line that forms the rectangular shape is now an individual object.

EXTEND

The **EXTEND** command is used to extend an object to a **boundary.** The object to be extended must actually or theoretically intersect the boundary.

1. Select the **EXTEND** command using one of the following:

   Ribbon = Home tab / Modify panel / ⊣√

   Keyboard = EX <enter>

2. The following will appear on the command line:

Command: _extend
Current settings: Projection = UCS   Edge = Extend
Select boundary edges ...
Select objects or <select all>: ***select boundary (P1) by clicking on the object.***
Select objects: ***stop selecting boundaries by selecting <enter>.***
Select object to extend or shift-select to Trim or
[Fence/Crossing/Project/Edge/Undo]:***select the object that you want to extend (P2 and P3). (Select the end of the object that you want to extend.)***
Select object to extend or [Fence/Crossing/Project/Edge/Undo]:***stop selecting objects by pressing <enter>.***

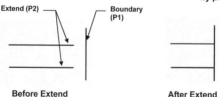

Extend (P2)   Boundary (P1)

Before Extend   **After Extend**

Note: When selecting the object to be extended (P2 and P3 above) click on the end pointing towards the boundary.

Fence Use a "Fence" line to select objects to extend
Crossing You may select objects using a Crossing Window
Project Same as Edge except used only in "3D".

Edge (Extend or No Extend)
   In the **"Extend"** mode, (default mode) the boundary and the Objects to be extended need only intersect if the objects were infinite in length.
   In the **"No Extend"** mode the boundary and the objects to be extended must visibly intersect.

Undo You may "undo" the last extended object while in the Extend command

FILLET

The **FILLET** command will create a radius between two objects. The objects do not have to be touching. If two parallel lines are selected, it will construct a full radius.

RADIUS A CORNER

1. Select the **FILLET** command using one of the following:

    Ribbon = Home tab / Modify panel /

    Keyboard = F <enter>

2. The following will appear on the command line:

3. **Set the radius of the fillet**
Command: _fillet
Current settings: Mode = TRIM, Radius = 0.000
Select first object or [Undo/Polyline/Radius/Trim/Multiple]:   *type "R" <enter>*
Specify fillet radius <0.000>: *type the radius <enter>*

4. **Now fillet the objects**
Select first object or [Undo/Polyline/Radius/Trim/Multiple]: *select the 1st object to be*
filleted
Select second object or shift-select to apply corner: *select the 2nd object to be filleted*

Polyline: This option allows you to fillet all intersections of a Polyline in one operation, such as all 4 corners of a rectangle.

Trim: This option controls whether the original lines are trimmed to the end of the Arc or remain the original length. (Set to Trim or No trim)

Multiple: Repeats the fillet command until you press <enter> or esc key.

1-18

FILLET....continued

The FILLET command may also be used to create a square corner.

SQUARE CORNER

1. Select the **FILLET** command using one of the following:

2. The following will appear on the command line:

3. **Select the two lines to form the square corner**
Select first object or [Undo/Polyline/Radius/Trim/Multiple]: *select the 1st object (P1)*

Select second object or shift-select to apply corner: *Hold the shift key down while selecting the 2nd object (P2)*

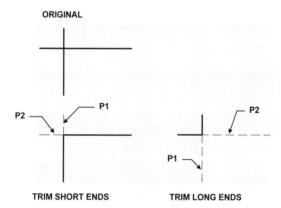

ORIGINAL

TRIM SHORT ENDS   TRIM LONG ENDS

Note: The corner trim direction depends on which end of the object you select.

MATCH PROPERTIES

Match Properties is used to "paint" the properties of one object to another. This is a simple and useful command. You first select the object that has the desired properties (the source object) and then select the object you want to "paint" the properties to (destination object).

Only one "source object" can be selected but its properties can be painted to any number of "destination objects".

1. Select the Match Properties command using one of the following:

   Ribbon = Home tab / Clipboard panel /

   Keyboard = MA <enter>

   Command: matchprop
2. Select source object: *select the object with the desired properties to match*

3. Select destination object(s) or [Settings]: *select the object(s) you want to receive the matching properties.*

4. Select destination object(s) or [Settings]: *select more objects or <enter> to stop.*

Note: If you do not want to match all of the properties, after you have selected the source object, right click and select "Settings" from the short cut menu, before selecting the destination object. Uncheck all the properties you do not want to match and select the OK button. Then select the destination object(s).

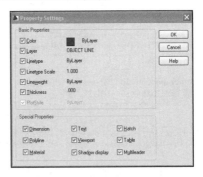

MEASURE COMMAND

The **MEASURE** command is very similar to the **DIVIDE** command because point objects are drawn at intervals on an object. However, the **MEASURE** command allows you to specify the **LENGTH** of the segments rather than the number of segments.

Note: the object selected is **NOT** broken into segments. The **POINTS** are simply drawn **ON** the object.

EXAMPLE:

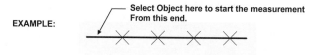

The **MEASURE**ment was started at the left endpoint, and ended just short of the right end of the line. The remainder is less than the measurement length specified.
You designate which end you want the measurement to start by selecting the end when prompted to select the object.

1. First open the Point Style box and select the **POINT STYLE** to be placed on the object.

   Keyboard = ddptype <enter>

(Refer to the Point command if you need a refresher on Points)

2. Next select the **MEASURE** command using one of the following:

   Ribbon = Home tab / Draw panel ▾ / [icon] (on the Point tool)

   Keyboard = ME <enter>

3. Select Object to MEASURE: *select the object to MEASURE*
   (Note: this selection point is also where the MEASUREment will start.)

4. Specify length of segment or [Block]: *type the length of one segment <enter>*

MEASURE TOOLS and ID Point

The following tools are very useful to confirm the location or size of objects.

The Measure tools enables you to measure the distance, radius, angle area, or volume of a selected object. The default option is Distance.

1..You may access these tools as follows:

Ribbon = Home tab / Utilities Panel / Distance ▾

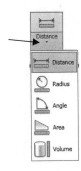

2. Select one of the tools and follow the instructions on the command line.

ID Point
The ID Point command will list the X and Y coordinates of the point that you select. The coordinates listed will be from the Origin.

1. Select the ID Point command by typing: **ID <enter>**
2. Select a location point, such as the endpoint of a line.
    The X and Y coordinates of the endpoint will be displayed.

Example:
1. Command: **id<enter>**
2. Snap to the endpoint
3. Coordinates, from the Origin, are displayed.

    2. Snap to
      endpoint

    Specify point: X = 6.8293   Y = 6.6926   Z = 0.0000

MIRROR

The **MIRROR** command allows you to make a mirrored image of any objects you select. You can use this command for creating right / left hand parts or draw half of a symmetrical object and mirror it to save drawing time.

1. Select the **MIRROR** command using one of the following:

   Ribbon = Home tab / Modify panel /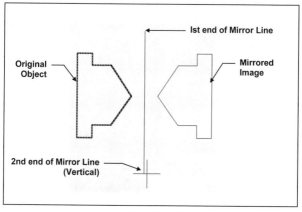

   Keyboard = MI <enter>

2. The following will appear on the command line:

Command: _mirror
Select objects: ***select the objects to be mirrored***
Select objects: ***stop selecting objects by selecting <enter>***
Specify first point of mirror line: ***select the 1st end of the mirror line***
Specify second point of mirror line: ***select the 2nd end of the mirror line***
Erase source objects? [Yes/No] <N>: ***select Y or N***

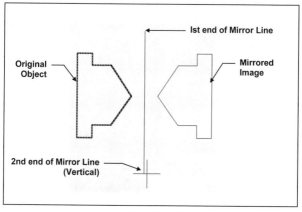

Mirror Line "Vertical"

Continued on the next page...

MIRROR....continued

Note:
The placement of the "Mirror Line" is important. You may make a mirrored copy
horizontally, vertically or on an angle. See examples below and on the previous page.

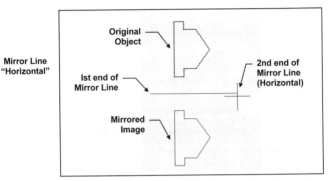

**Mirror Line
"Horizontal"**

Original Object

1st end of Mirror Line

2nd end of Mirror Line (Horizontal)

Mirrored Image

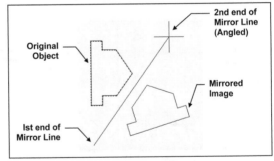

**Mirror Line
"Angled"**

2nd end of Mirror Line (Angled)

Original Object

Mirrored Image

1st end of Mirror Line

How to control text when using the Mirror command:
(Do the following before you use the mirror command)

1. At the command line type   **mirrtext <enter>**
2. If you want the text to mirror: type 1 <enter>
   If you do not want the text to mirror: type 0 <enter>

MIRRTEXT SETTING = 1   MIRRTEXT SETTING = 0

MOVE

The **MOVE** command is used to move object(s) from their current location (base point) to a new location (second displacement point).

1. Select the Move command using one of the following:

    Ribbon = Home tab / Modify panel / [icon]

    Keyboard = M <enter>

2. The following will appear on the command line:

Command: _move
Select objects: **select the object(s) you want to move (P1).**
Select objects: **select more objects or stop selecting object(s) by selecting <enter>.**
Specify base point or displacement: **select a location (P2) (usually on the object).**
Specify second point of displacement or <use first point as displacement>: **move the object to its <u>new location</u> (P3) and press the left mouse button.**

Warning: If you press <enter> instead of actually picking a new location (P3) with the mouse, Autocad will send it into <u>Outer Space</u>. If this happens just select undo tool and try again.

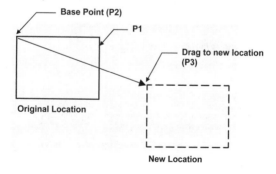

Base Point (P2)

P1

Drag to new location (P3)

Original Location

New Location

OFFSET

The **OFFSET** command duplicates an object parallel to the original object at a specified distance. You can offset Lines, Arcs, Circles, Ellipses, 2D Polylines and Splines. You may duplicate the original object or assign the offset copy to another layer.

Examples of Offset objects.

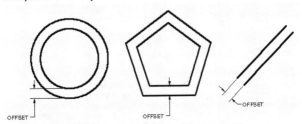

HOW TO USE THE OFFSET COMMAND:

METHOD 1
(Duplicate the Original Object)

1. Select the **OFFSET** command using one of the following:

   Ribbon = Home tab / Modify panel /

   Keyboard = Offset <enter>

2. Command: _offset
   Current settings: Erase source=No Layer=Source OFFSETGAPTYPE=0
   Specify offset distance or [Through/Erase/Layer] <Through>: *type the offset distance or select Erase or Layer. (see options on the next page)*

3. Select object to offset or <Exit/Undo>: *select the object to offset.*

4. Specify point on side to offset or [Exit/Multiple/Undo]<Exit>: *Select which side of the original you want the duplicate to appear by placing your cursor and clicking. (See options on the next page)*

5. Select object to offset or [Exit/Undo]<Exit>: *Press <enter> to stop.*

OFFSET....continued

METHOD 2
(Duplicate original object but assign the Offset copy to a different layer)

To automatically place the <u>offset copy</u> on a different layer than the original you must first change the "current" layer to the layer you want the offset copy to be placed on.

1. Select the layer that you want the offset copy placed on from the list of layers.

2. Select the OFFSET command (refer to previous page)

3. Command: _offset
   Current settings: Erase source=No Layer=Source OFFSETGAPTYPE=0
   Specify offset distance or [Through/Erase/Layer] <Through>: *type L <enter>*

4. Enter layer option for offset objects [Current/Source] <Source>: *select C <enter>*

5. Specify offset distance or [Through/Erase/Layer] <Through>: *type the offset dist*
                                                                <enter>
6. Select object to offset or [Exit/Undo] <Exit>: *select the object (source) to offset.*

7. Specify point on side to offset or [Exit/Multiple/Undo]<Exit>: *Select which side of the object you want the duplicate to appear by placing your cursor and clicking. (See options below)*

8. Select object to offset or [Exit/Undo]<Exit>: *Press <enter> to stop.*

OPTIONS:
Through: Creates an object passing through a specified point.

Erase: Erases the source object after it is offset.

Layer: Determines whether offset objects are created on the <u>current</u> layer or on the layer of the <u>source</u> object. Select <u>Layer</u> and then select <u>current</u> or <u>source.</u>
(Source is the default)

Multiple: Turns on the multiple offset mode, which allows you to continue creating duplicates of the original without re-selecting the original.

Exit: Exits the Offset command.

Undo: Removes the previous offset copy.

ROTATE

The **ROTATE** command is used to rotate objects around a Base Point. (pivot point)

After selecting the objects and the base point, you will enter the rotation angle from its <u>current</u> rotation angle or select a reference angle followed by the new angle.

A **Positive** rotation angle revolves the objects **Counter- Clockwise**.
A **Negative** rotation angle revolves the objects **Clockwise**.

Select the ROTATE command using one of the following:

Ribbon = Home tab / Modify panel /

Keyboard = RO <enter>

ROTATION ANGLE OPTION
Command: _rotate
1. Current positive angle in UCS: ANGDIR=counterclockwise ANGBASE=0
   Select objects: *select the object to rotate.*
2. Select objects: *select more object(s) or <enter> to stop.*
3. Specify base point: *select the base point (pivot point).*
4. Specify rotation angle or [Copy/Reference]<0>: *type the angle of rotation.*

REFERENCE OPTION
Command: _rotate
1. Current positive angle in UCS: ANGDIR=counterclockwise ANGBASE=0
   Select objects: *select the object to rotate.*
2. Select objects: *select more object(s) or <enter> to stop.*
3. Specify base point: *select the base point (pivot point).*
4. Specify rotation angle or [Reference]: *select Reference.*
5. Specify the reference angle <0>: *Snap to the reference object (1) and (2).*
6. Specify the new angle or [Points]: *P <enter>.*
7. Specify first point: *select 1st endpoint of new angle*
8. Specify second point: *select 2nd endpoint of new angle*

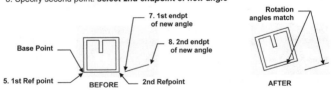

SCALE

The **SCALE** command is used to make objects larger or smaller <u>proportionately</u>. You may scale using a scale factor or a reference length. You must also specify a base point. The base point is a stationary point from which the objects scale.

1. Select the SCALE command using one of the following:

   Ribbon = Home tab / Modify panel / 🔲

   Keyboard = SCALE <enter>

SCALE FACTOR

   Command: _scale
2. Select objects: *select the object(s) to be scaled*
3. Select objects: *select more object(s) or <enter> to stop*
4. Specify base point: *select the stationary point on the object*
5. Specify scale factor or [Copy/Reference]: *type the <u>scale factor</u> <enter>*

If the scale factor is greater than 1, the objects will increase in size.
If the scale factor is less than 1, the objects will decrease in size.

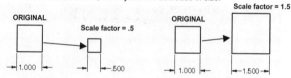

REFERENCE option

Command: _scale
2. Select objects: *select the object(s) to be scaled*
3. Select objects: *select more object(s) or <enter> to stop*
4. Specify base point: *select the stationary point on the object*
5. Specify scale factor or [Copy/Reference]: *select Reference*
6. Specify reference length <1>: *specify a <u>reference</u> length*
7. Specify new length: *specify the new length*

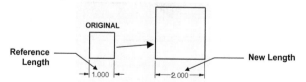

<u>COPY</u> option - creates a duplicate of the selected object. The duplicate is directly on top of the original. The duplicate will be scaled. The Original remains the same.

STRETCH

The **STRETCH** command allows you to stretch or compress object(s). Unlike the Scale command, you can alter an objects proportion with the Stretch command. In other words, you may increase the length without changing the width and vice versa.

Stretch is a very valuable tool. Take some time to really understand this command. It will save you hours when making corrections to drawings.

When selecting the object(s) you must use a **CROSSING** window.
Objects that are crossed, will *stretch.*
Objects that are totally enclosed, will *move*.

1. Select the STRETCH command using one of the following:

   Ribbon = Home tab / Modify panel /

   Keyboard = S <enter>

   Command: _stretch
2. Select objects to stretch by crossing-window or crossing-polygon...
   Select objects: *select the first corner of the crossing window*
3. Specify opposite corner: *specify the opposite corner of the crossing window*
4. Select objects: *<enter>*
5. Specify base point or [Displacement] <Displacement>:
   select a base point (where it stretches from)
6. Specify second point or <use first point as displacement>:
   type coordinates or place location with cursor

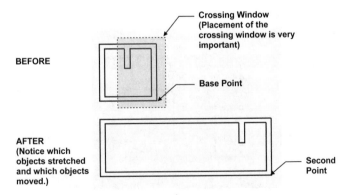

BEFORE

Crossing Window
(Placement of the crossing window is very important)

Base Point

AFTER
(Notice which objects stretched and which objects moved.)

Second Point

TRIM

The **TRIM** command is used to trim an object to a **cutting edge**. You first select the "Cutting Edge" and then select the part of the object you want to trim. The object to be trimmed must actually intersect the cutting edge or <u>could</u> intersect if the objects were infinite in length.

1. Select the Trim command using one of the following:

   Ribbon = Home tab / Modify panel / ⊬

   Keyboard = TR <enter>

2. The following will appear on the command line:

Command: _trim
Current settings: Projection = UCS   Edge = Extend
Select cutting edges ...
Select objects or <select all>: ***select cutting edge(s) by clicking on the object (P1)***
Select objects: ***stop selecting cutting edges by pressing the <enter> key***
Select object to trim or shift-select to extend or
[Fence/Crossing/Project/Edge/eRase/Undo]: ***select the object that you want to trim. (P2)
(Select the part of the object that you want to disappear, not the part you want to remain)***
Select object to trim or [Fence/Crossing/Project/Edge/eRase/Undo]: ***press <enter> to stop***

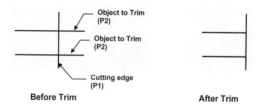

Before Trim   **After Trim**

Note: You may toggle between Trim and Extend. Hold down the shift key and the Extend command will activate. Release the shift key and you return to Trim.

Fence Use a "Fence" line to select objects to extend
Edge See page 1-17
Project Same as Edge except used only in "3D".

Crossing You may select objects using a Crossing Window.

eRase You may erase an object instead of trimming while in the Trim command.

Undo You may "undo" the last trimmed object while in the Trim command

UNDO and REDO

The **UNDO** and **REDO** tools allow you to undo or redo previous commands.
For example, if you erase an object by mistake, you can UNDO the previous "erase" command and the object will reappear. So don't panic if you do something wrong. Just use the UNDO command to remove the previous commands.

The **Undo** and **Redo** tools are located in the **Quick Access Toolbar**.

Note:
You may UNDO commands used during a work session until you close the drawing.

How to use the <u>Undo</u> tool.

1. Start a new drawing.
2. Draw a line, circle and a rectangle.

Your drawing should look approximately like this.

3. Next Erase the Circle and the Rectangle.

(The Circle and the Rectangle disappear.)

4. Select the UNDO arrow.

You have now deleted the ERASE command operation.
As a result the erased objects reappear.

How to use the <u>Redo</u> command:
Select the REDO arrow and the Circle and Rectangle will disappear again.

WIPEOUT

The Wipeout command creates a blank area that covers existing objects. The area has a background that matches the background of the drawing area. This area is bounded by the wipeout frame, which you can turn on or off.

1. Select the Wipeout command using one of the following:

   Ribbon = Home tab / Draw ▾ panel / [image]

   Keyboard = Wipeout <enter>

2. Command: _wipeout Specify first point or [Frames/Polyline] <Polyline>: *specify the first point of the shape (P1)*
3. Specify next point: *specify the next point (P2)*
4. Specify next point or [Undo]: *specify the next point (P3)*
5. Specify next point or [Undo]: *specify the next point or <enter> to stop*

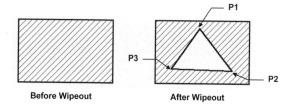

Before Wipeout   **After Wipeout**

TURNING FRAMES ON OR OFF

1. Select the Wipeout command.
2. Select the "Frames" option.
3. Enter ON or OFF.

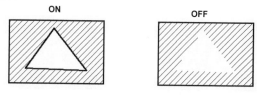

ON   OFF

Note: If you want to move the objects and the wipeout area, you must select both and move them at the same time. Do not move them separately.

ZOOM

The **ZOOM** command is used to move closer to or farther away from an object.
This is called Zooming In and Out.

1. Select the Zoom command by using the following:

   Ribbon = View tab / Navigate panel

2. Select the ⌄ down arrow to display all of the selections.

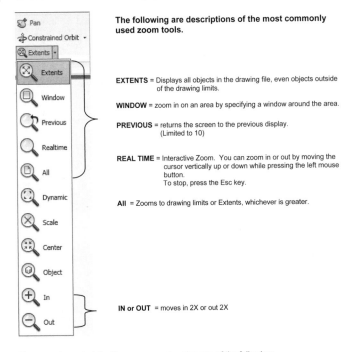

The following are descriptions of the most commonly used zoom tools.

EXTENTS = Displays all objects in the drawing file, even objects outside of the drawing limits.

WINDOW = zoom in on an area by specifying a window around the area.

PREVIOUS = returns the screen to the previous display. (Limited to 10)

REAL TIME = Interactive Zoom. You can zoom in or out by moving the cursor vertically up or down while pressing the left mouse button.
To stop, press the Esc key.

All = Zooms to drawing limits or Extents, whichever is greater.

IN or OUT = moves in 2X or out 2X

You may also select the Zoom commands using one of the following:

Right Click and select Zoom from the Short cut menu.

Keyboard = Z <enter> Select from the options listed..

ZOOM....continued

How to use ZOOM / WINDOW

1. Select Zoom / Window (Refer to previous page)

2. Draw a window around the objects you want to enlarge.
   (Drawing a "window" is the similar to drawing a rectangle. It requires a first corner
   and then diagonal corner)

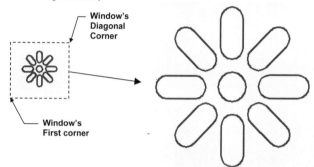

Magnified to this view
Note: the objects have been magnified.
But the size has not changed.

How to return to Original View

1. Type: **Z** <enter> **A** <enter> (This is a shortcut for Zoom / All)

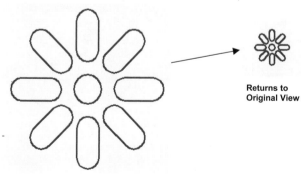

**Returns to
Original View**

Notes

Section 2
Concepts

MODEL and LAYOUT OPTIONS

Very important:
*Before I discuss Model and Layout I need you to make a change to the options. The default AutoCAD screen does not display the **Model and Layout tabs**. AutoCAD gives you the option to displayed them or not. I prefer to have the tabs displayed. I think it will make it easier for you to understand.*

This will just take a minute.

1. Type **options <enter>**

2. Select the **Display** tab.

3. Check and un-check boxes as shown

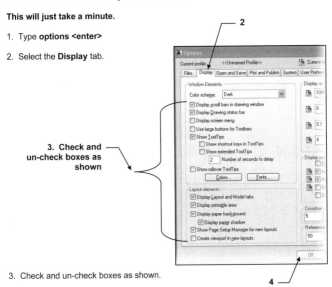

3. Check and un-check boxes as shown.

4. Select the **OK** button

5. The lower left corner of the drawing area should now display the 3 tabs Model, Layout1 and Layout2 and a few new tools should be displayed in the lower right corner above the command line.

5. New

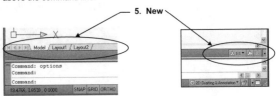

MODEL and LAYOUT tabs

AutoCAD provides two drawing spaces, **MODEL** and **LAYOUT**. You move into one or the other by selecting either the MODEL or LAYOUT tabs, located at the bottom left of the drawing area. (If you do not have these displayed follow the instructions on the previous page.)

Refer to the previous page if you do not have these.

Model Tab (Also called *Model Space*)

When you select the Model tab you enter <u>MODEL SPACE</u>.

Model Space is where you **create** and **modify** your drawings.

Layout Tabs (Also called *Paper Space*)

When you select a Layout tab you enter <u>PAPER SPACE</u>.
<u>The primary function of Paper Space is to **prepare the drawing for plotting.**</u>

When you select the Layout tab for the first time, the "<u>Page Setup Manager</u>" dialog box will appear. The Page Setup Manager allows you to specify the printing device and paper size to use.
(More information on this in "How to ......)

When you select a Layout tab, Model Space will seem to have disappeared, and a <u>blank sheet of paper</u> is displayed on the screen. This sheet of paper is basically <u>in front of the Model Space</u>. (Refer to the illustration on the next page)

To see the drawing in Model Space, while still in Paper Space, you must <u>cut a hole</u> in this sheet. This hole is called a **"Viewport"**. *(Refer to "How to ....)*

MODEL and LAYOUT tabs....continued

Try to think of this as a picture frame (paper space) in front of a photograph (model space).

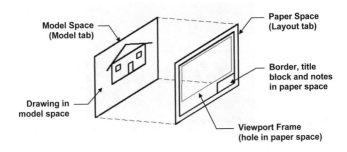

Model Space
(Model tab)

Paper Space
(Layout tab)

Border, title
block and notes
in paper space

Drawing in
model space

Viewport Frame
(hole in paper space)

This is what you see when you
select the ▶ Model ◀ tab.

You see only model space.

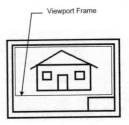

Viewport Frame

This is what you see when you
select the Layout1 tab with
a viewport.

You see through the Viewport to
Model Space.

WHY LAYOUTS ARE USEFUL

A Layout (Paper Space) is a great method to manipulate your drawing for plotting.

Notice the drawing below with <u>multiple viewports</u>.

Each viewport is a hole in the paper.
You can see through each viewport (hole) to model space.

Using Zoom and Pan you can manipulate the display of model space in each viewport.
To manipulate the display you must be <u>inside</u> the viewport.

Note: the dashed line indicates the maximum printing area for the printer and paper selected. Any object outside of this area will not print.

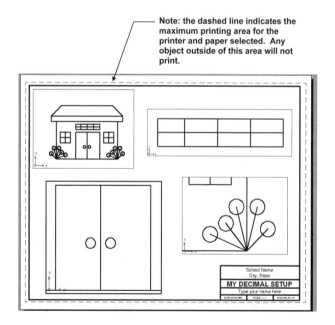

CREATING SCALED DRAWINGS

A very important rule in CAD you must understand is:

"All objects are drawn full size".

In other words, if you want to draw a line 20 feet long, you actually draw it 20 feet long. If the line is 1/8" long, you actually draw it 1/8" long.

Drawing and Plotting objects that are very large or very small.

What if you wanted to draw a house? Could you print it to scale on an 8-1/2 X 11 piece of paper? How about a small paper clip. Could you make it big enough to dimension?

How to print an entire house on an 8-1/2 X 11 sheet of paper.

Remember the photo and picture frame example I suggested on page 2-4. This time try to picture yourself standing at the front door of your house with an empty picture frame in your hands. Look at your house through the picture frame. Of course the house is way too big to fit in the frame. Or is it because you are standing too close to the house?

Now walk across the street and look through the picture frame in your hands again. Does the house appear smaller? Can you see all of it in the frame? If you could walk far enough away from the house it would eventually appear small enough to fit in the picture frame in your hands. But....the house did not actually change size, did it? It only appears smaller because you and the picture frame are farther away from it.

Adjusting the Viewport scale.

When using AutoCAD, walking across the street with the frame in your hands is called **Adjusting the Viewport scale.** You are increasing the distance between model space (your drawing) and Paper space (Layout) and that makes the drawing appear smaller.

For example: A viewport scale of 1/4" = 1' would make model space appear 48 times smaller. But, when you dimension the house, the dimension values will be the actual measurement of the house. In other words, a 30 ft. line will have a dimension of 30'-0".

When plotting something smaller, like a paperclip, you have to move the picture frame closer to model space to make it appear larger. For example: 8 = 1   .

ADJUSTING THE VIEWPORT SCALE

The following will take you through the process of adjusting the scale within a viewport.

1. **Open** a drawing.

2. Select a **Layout** tab. (paper space)

3. Cut a new Viewport or unlock an existing Viewport. (See section "How to....")

4. **Zoom / All** to display all of the drawing limits.

5. Adjust the scale
   A. You must be in Paper Space.
   B. Select the Viewport Frame.
   C. Unlock Viewport, if locked.
   D. Select the Viewport Scale down arrow.
   E. Select the scale from the list of scales.

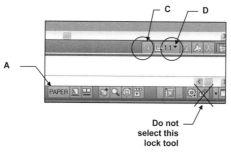

Do not select this lock tool

6. Lock the Viewport

Note: If you would like to add a scale that is not on the list:
   1. Type **Scalelistedit**
   2. Select **Add** button.
   3. Enter Scale name to display in scale list.
   4. Enter Paper and drawing units.
   5. Select the **OK** button.

Notes

Section 3
Dimensioning

DIMENSIONING

Dimensions can be Associative, Non-Associative or Exploded. You need to understand what these are so you may decide which setting you want to use. Most of the time you will use Associative but you may have reasons to also use Non-Associative and Exploded.

Associative
Associative Dimensioning means that the dimension is actually associated to the objects that they dimension. If you move the object, the dimension will move with it. If you change the size of the object, the dimension text value will change also.
(Note: This is not parametric. In other words, you cannot change the dimension text value and expect the object to change. That would be parametric dimensioning)

Non-Associative
Non-Associative means the dimension is not associated to the objects and will not change if the size of the object changes.

Exploded
Exploded means the dimension will be exploded into lines, text and arrowheads and non-associative.

How to set dimensioning to Associative, Non-Associative or Exploded.

1. On the command line type: ***dimassoc <enter>***

2. Enter the number ***2, 1 or 0 <enter>.***

    2 = Associative

    1 = Non-Associative

    0 = Exploded

DIMENSIONING....continued

How to Re-associate a dimension.
If a dimension is <u>Non-associative</u>, and you would like to make it <u>Associative</u>, you may use the **dimreassociate** command to change.

1. Select **Reassociate** using one of the following:

   Ribbon = Annotate tab / Dimension ▾ panel /

   Keyboard = Dimreassociate <enter>

2. Select objects: *select the dimension to be reassociated.*

3. Select objects: *select more dimensions or <enter> to stop.*

4. Specify first extension line origin or [Select object] <next>: (*an "X" will appear to identify which is the first extension); use object snap to select the exact location, on the object, for the extension line point.*

5. Specify second extension line origin <next>: (*the "X" will appear on the second extension) use object snap to select the exact location, on the object, for the extension line point.*

6. *Continue until all extension line points are selected.*

<u>Note: You must use object snap to specify the exact location for the extension lines.</u>

Regenerating Associative dimensions
Sometimes after panning and zooming, the associative dimensions seem to be floating or not following the object. The **DIMREGEN** command will move the associative dimensions back into their correct location.

Type: dimregen <enter>

DIMENSION STYLES

Using the "Dimension Style Manager" you can change the appearance of the dimension features, such as length of arrowheads, size of the dimension text, etc. There are over 70 different settings.
You can also Create New, Modify and Override Dimension Styles. All of these are simple, by using the Dimension Style Manager described below.

1. Select the "Dimension Style Manager" using one of the following:

   Ribbon = Annotate tab / Dimension panel / ↘

   Keyboard = Dimstyle <enter>

Displays which style is currently in use. ⟶

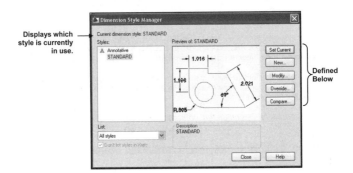

Defined Below

Set Current Select a style from the list of styles and select the **set current** button.

New Select this button to create a new style. When you select this button, the **Create New Dimension Style** dialog box is displayed.

Modify Selecting this button opens the **Modify dimension Style** dialog box which allows you to make changes to the style selected from the "list of styles".

Override An override is a temporary change to the current style. Selecting this button opens the Override Current Style dialog box.

Compare Compares two styles.

CREATING A NEW DIMENSION STYLE

A dimension style is a group of settings that has been saved with a name you assign.
When creating a new style you must start with an existing style, such as Standard.
Next, assign it a new name, make the desired changes and when you select the OK
button the new style will have been successfully created.

How to create a NEW dimension style.

1. Open a drawing.

2. Select the **Dimension Style Manager** command (Refer to previous page)

3. Select the **NEW** button.

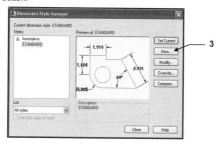

4. Enter **CLASS STYLE** in the "**New Style Name**" box.

5. Select **STANDARD** in the "**Start With:**" box.

6. "**Use For:**" box will be discussed later. For now, leave it set to "**All dimensions**".

7. Select the **CONTINUE** button.

CREATING A NEW DIMENSION STYLE....continued

8. Select the **Primary Units** tab and change your setting to match the settings shown below.

8. Primary Units

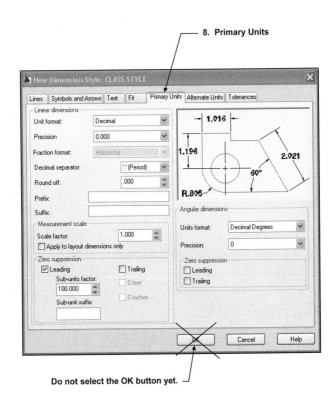

Do not select the OK button yet.

CREATING A NEW DIMENSION STYLE....continued

9. Select the **Lines** tab and change your settings to match the settings shown below.

9. Lines

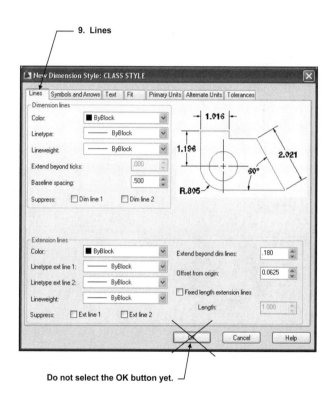

Do not select the OK button yet.

CREATING A NEW DIMENSION STYLE....continued

10. Select the **Symbols and Arrows** tab and change your setting to match the settings shown below.

10. Symbols and Arrows

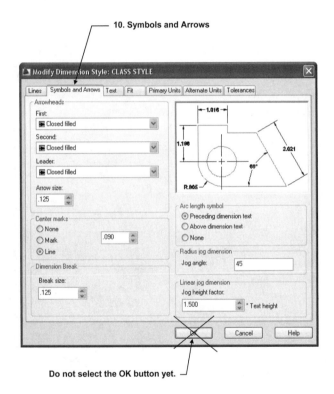

Do not select the OK button yet.

CREATING A NEW DIMENSION STYLE....continued

11. Select the **Text** tab and change your settings to match the settings shown below.

11. Text

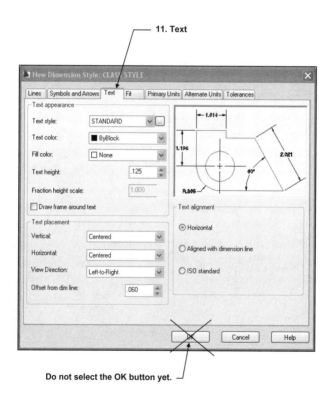

Do not select the OK button yet.

CREATING A NEW DIMENSION STYLE....continued

12. Select the **Fit** tab and change your setting to match the settings shown below.

13. Now select the OK button.

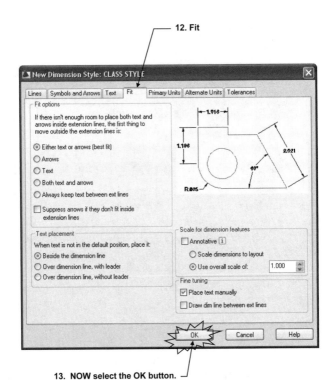

12. Fit

13. NOW select the OK button.

CREATING A NEW DIMENSION STYLE....continued

Your new style **"Class Style"** should be listed.

14. Select the **"Set Current"** button to make your new style "Class Style" the style that will be used.

14 ⎯⎯⎯⎯⎯⎯⎯⎯

⎯ **New Style listed here**

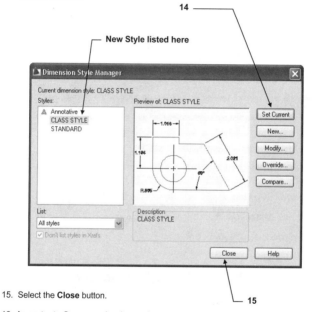

15. Select the **Close** button.   � 15

16. **Important:** Save your drawing again.

Note: You have successfully created a new "_Dimension_ Style" called **"Class Style"**.
This style will be saved _in your drawing_ after you save the drawing.
It is important that you understand that this dimension style resides _only_ in this
drawing. If you open another drawing, this dimension style will not be there.

CREATING A DIMENSION SUB-STYLE

Previously you learned how to create a Dimension Style named Class Style. All of the dimension created with that style appear identical because they have the same settings. Now you are going to learn how to create a "Sub-Style" of the Class Style.

For example:
If you wanted all of the **Diameter** dimensions to have a centerline automatically displayed but you did not want a centerline displayed when using the **Radius** dimension command. To achieve this, you must create a "sub-style" for all **Radius** dimensions.

Sub-styles have also been called "children" of the "Parent" dimension style. As a result, they form a family.

A Sub-style is permanent, unlike the Override command, which is temporary.

Note: This sounds much more complicated than it is. Just follow the steps below. It is very easy.

How to create a sub-style for Radius dimensions.

You will set the center mark to None for the Radius command only.
The Diameter command center mark will not change.

1. Open your drawing.

2. Select the **DIMENSION / STYLE** command. (Refer to page 16-8)

3. Select **"Class Style"** from the Style List.

4. Select the **NEW** button.

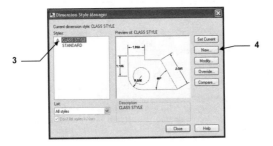

Continued on the next page...

EDITING DIMENSION TEXT VALUES

Sometimes you need to modify the dimension text value. You may add a symbol, a note or even change the text of an existing dimension.
The following describes 2 methods.

Example: Add the word "Max." to the existing dimension value text.

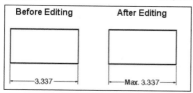

Before Editing   After Editing

─ 3.337 ─   ─ Max. 3.337 ─

Method 1 (Properties Palette).
1. **Quick Properties** button must be **ON**.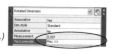
2. Select the dimension you want to override.
3. Scroll down to **Text override**
   (Notice that the actual <u>measurement</u> is directly above it.)
4. Type the new text (**Max**) and **< >** and press <enter>
   (**< >** represents the associative text value 3.337)

Important:
If you do not use < > the dimension will no longer be Associative.

Method 2 (Text Edit).
1. Type on the Command line: **ed** <enter>   (This is the **text edit** command)
2. Select the dimension you want to edit.

   Associative Dimension
   If the dimension is <u>Associative</u> the dimension text will appear highlighted.

    ─ **Before**

   You may add text in front or behind the dimension text and it will remain
   Associative. Be careful not to disturb the dimension value text.

   ─ Max 9.815 ─ **After**

   Non Associative or Exploded Dimension
   If the dimension value has been changed or exploded it will appear with a gray
   background and is not Associative.

   ─ 9.815 Max ─

4. Make the change.
5. Select the **OK** button.

EDITING THE DIMENSION POSITION

Sometimes dimensions are too close and you would like to stagger the text or you
need to move an entire dimension to a new location, such as the examples below.

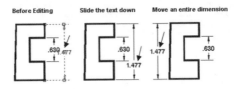

Before Editing   **Slide the text down**   **Move an entire dimension**

Editing the position is easy.
1. Select the dimension that you want to change. (Grips will appear)
2. Select the middle grip. (It will become red and active, hot)
3. Move the cursor and the dimension will respond. (Osnap should be OFF)
4. Press the left mouse button to place the dimension in the new location.

(Refer to "grips" in Section 8)

ADDITIONAL EDITING OPTIONS USING THE SHORTCUT MENU

1. Select the dimension that you want to change.
2. Press the Right Mouse button.
3. Select "Dimension Text Position" from the shortcut menu shown to the right.
4. A sub-menu appears with many editing options.

MODIFY AN <u>ENTIRE</u> DIMENSION STYLE

After you have created a Dimension Style, you may find that you have changed your mind about some of the settings. You can easily change the entire Style by using the "Modify" button in the Dimension style Manager dialog box. This will not only change the Style for future use, but it will also <u>update dimensions already in the drawing</u>.

Note: if you do not want to update the dimensions already in the drawing, but want to make a change to a new dimension, refer to <u>Override</u>.

1. Select the <u>Dimension Style Manager</u>.

2. Select the Dimension Style that you wish to modify.

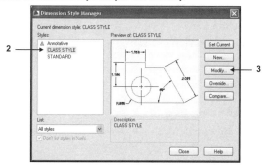

3. Select the **Modify button** from the Dimension Style Manager dialog box.

4. Make the desired changes to the settings.

5. Select the **OK** button.

6. Select the **Close** button.

Now look at your drawing. Have your dimensions updated?

Note:
The method above <u>will not</u> change dimensions that have been modified or exploded.

<u>Note: If some of the dimensions have not changed:</u>
1. Type: **-Dimstyle <enter>**   (notice the (-)dash infront of "dimstyle")
2. Type: **A <enter>**
3. Select dimensions to update and then <enter>
(Sometimes you have to give them a little nudge.)

OVERRIDE A DIMENSION STYLE

A dimension Override is a **temporary** change to the dimension settings.
An override **will not affect existing dimensions**. It will **only** affect **new dimensions**.
Use this option when you want a **new** dimension just a little bit different but you don't
want to create a whole new dimension style and you don't want the existing
dimensions to change either.

For example, if you want the new dimension to have a text height of .500 but you want
the existing text to remain at .125 ht.

1. Select the **Dimension Style Manager.**

2. Select the "**Style"** you want to override. (Such as: Class Style)

3. Select the **Override** button.

4. Make the desired changes to the settings. (Such as: Text ht = .500)

5. Select the **OK** button.

6. Confirm the Override
   Look at the List of styles.
   Under the Style name, a sub heading of **<style overrides>** should be displayed.
   The description box should display the style name and the override settings.

7. The description box should display the style name and the override settings.

8. Select the **Close** button.

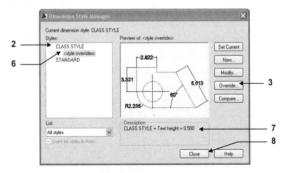

When you want to return to style **Class Style**, select **Class Style** from the styles list
then select the **Set Current** button. Each time you select a different style, you must
select the **Set Current** button to activate it.

EDIT AN INDIVIDUAL EXISTING DIMENSION

Sometimes you would like to modify the settings of an **individual existing** dimension. This can be achieved using the **Properties palette**.

1. Double click on the dimension that you wish to change.

2. Select and change the desired settings.

   Example: Change the dimension text height to .500.

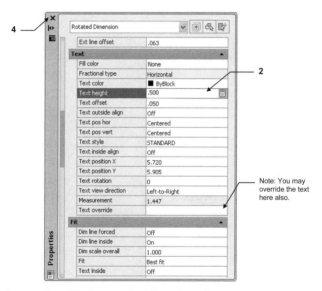

3. Press <enter> . *(The dimension should have changed)*

4. Close the Properties Palette.

5. Press the <esc> key to de-activate the grips.

Note: The dimension will remain Associative.

LINEAR DIMENSIONING

__Linear dimensioning__ allows you to create horizontal and vertical dimensions.

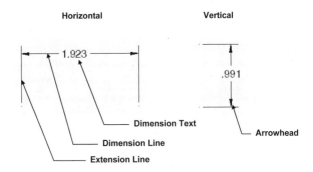

Horizontal   **Vertical**

──────── Dimension Text

──── Dimension Line

── Extension Line

─ Arrowhead

1. Select the **LINEAR** command using one of the following:

   Ribbon = Annotate tab / Dimension panel / [icon]

   Keyboard = Dimlinear <enter>

   Command:__dimlinear
2. Specify first extension line origin or <select object>: *snap to first extension line origin (P1).*
3. Specify second extension line origin: *snap to second extension line origin (P2).*

4. Specify dimension line location or [Mtext/Text/Angle/Horizontal/Vertical/Rotated]: *select where you want the dimension line placed (P3).*

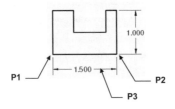

BASELINE DIMENSIONING

Baseline dimensioning allows you to establish a **baseline** for successive dimensions. The spacing between dimensions is automatic and should be set in the dimension style. (See Baseline spacing setting)

A Baseline dimension must be used with an existing dimension. If you use Baseline dimensioning immediately after a Linear dimension, you do not have to specify the baseline origin.

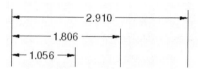

1. Create a <u>linear</u> dimension first **(P1 and P2).**

2. Select the **BASELINE** command using one of the following:
   Command: _dimbaseline

   Baseline is in the drop-down menu with Continue.

   Ribbon = Annotate tab / Dimension panel /

   Keyboard = Dimbaseline <enter>

3. Specify a second extension line origin or [Undo/Select] <Select>: *snap to the second extension line origin (P3).*

4. Specify a second extension line origin or [Undo/Select] <Select>: *snap to P4.*

5. Specify a second extension line origin or [Undo/Select} <Select>: *select <enter> twice to stop.*

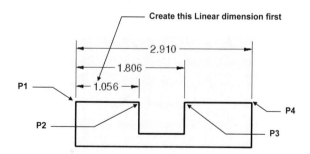

Create this Linear dimension first

CONTINUE DIMENSIONING

<u>Continue</u> creates a series of dimensions in-line with an existing dimension. If you use the continue dimensioning immediately after a Linear dimension, you do not have to specify the continue extension origin.

|← 1.056 →| .750 |← 1.105 →|

1. Create a linear dimension first (**P1 and P2 shown below**)

2. Select the Continue command using one of the following:

   Ribbon = Annotate tab / Dimension panel / |⊢⊦⊣|

   Keyboard = Dimcontinue <enter>

   Command: _dimcontinue
3. Specify a second extension line origin or [Undo/Select] <Select>: *snap to the*
   second extension line origin (P3).

4. Specify a second extension line origin or [Undo/Select] <Select>: *snap to the*
   second extension line origin (P4).

5. Specify a second extension line origin or [Undo/Select] <Select>: *press <enter>*
   twice to stop.

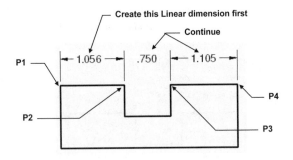

Create this Linear dimension first

Continue

P1

|← 1.056 →| .750 |← 1.105 →|

P4

P2

P3

ALIGNED DIMENSIONING

The **ALIGNED** dimension command aligns the dimension with the angle of the object that you are dimensioning. The process is the same as Linear dimensioning. It requires two extension line origins and the placement of the dimension text location. (Example below)

1. Select the **ALIGNED** command using one of the following:

   Ribbon = Annotate tab / Dimension panel /   Select ▾

   Keyboard = Dal <enter>

   Command: _dimaligned
2. Specify first extension line origin or <select object>: *select the first extension line origin (P1)*

3. Specify second extension line origin: *select the second extension line origin (P2)*

4. Specify dimension line location or [Mtext/Text/Angle]: *place dimension text location (P3)*

   Dimension text = *the dimension value will appear here*

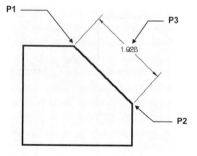

ANGULAR DIMENSIONING

The **ANGULAR** dimension command is used to create an angular dimension between two lines that form an angle. AutoCAD determines the angle between the selected lines and displays the dimension text followed by a degree (°) symbol.

1. Select the **ANGULAR** command.

   Ribbon = Annotate tab / Dimension panel /  **Select ▾**

   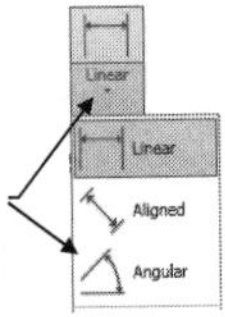

   Keyboard = Dimangular <enter>

2. Select arc, circle, line, or <specify vertex>: ***select the first line that forms the angle (P1) location is not important, <u>do not use object snap</u>.***

3. Select second line: ***select the second line that forms the angle (P2)***

4. Specify dimension arc line location or [Mtext/Text/Angle]: ***place dimension text location (P3)***

   Dimension text = ***angle will be displayed here***

   Any of the 4 angular dimensions shown below can be displayed by moving the cursor in the direction of the dimension after selecting the 2 lines (P1 and P2) that form the angle.

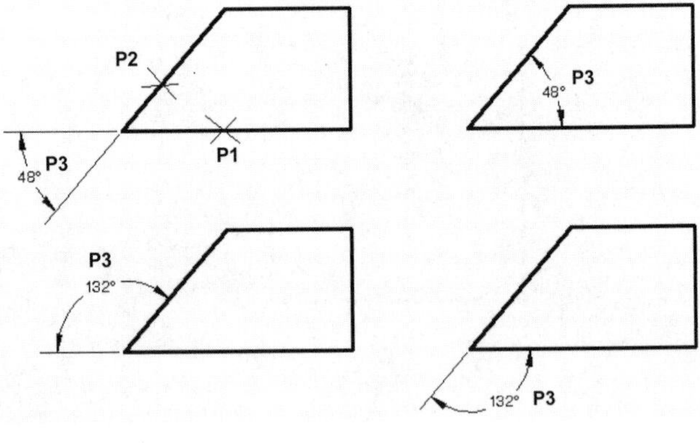

DIMENSIONING ARC LENGTHS

You may dimension the distance along an Arc. This is known as the **Arc length**.
Arc length is an <u>associative</u> dimension.

Example:

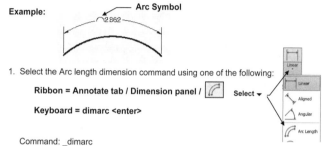

1. Select the Arc length dimension command using one of the following:

   Ribbon = Annotate tab / Dimension panel /

   Keyboard = dimarc <enter>

   Command: _dimarc
2. Select arc or polyline arc segment: **select the Arc**
3. Specify arc length dimension location, or [Mtext/Text/Angle/Partial/Leader]:
   place the dimension line and text location

   Dimension text = **dimension value will be shown here**

To differentiate the Arc length dimensions from Linear or Angular dimensions, arc length
dimensions display an arc (⌒) symbol by default. (Also called a "hat" or "cap")

The arc symbol may be displayed either <u>above,</u> or <u>preceding</u> the dimension text.
You may also choose not to display the arc symbol.

Example:

PRECEDING	ABOVE	NONE

Specify the placement of the arc symbol in the **Dimension Style / Symbols and Arrows**
tab or you may edit its position using the Properties Palette.

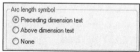

Continued on the next page...

DIMENSIONING ARC LENGTHS....continued

The extension lines of an Arc length dimension are displayed as <u>radial</u> if the included angle is <u>greater than 90 degrees</u>.

Example:

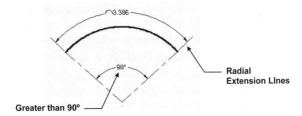

Radial
Extension LInes

Greater than 90°

The extension lines of an Arc length dimension are displayed as <u>orthogonal</u> if the included angle is <u>less than 90 degrees</u>.

Example:

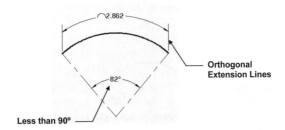

Orthogonal
Extension Lines

Less than 90°

DIMENSIONING A LARGE CURVE

When dimensioning an arc the dimension line should pass through the center of the arc. However, for large curves, the true center of the arc could be very far away, even off the sheet.

When the true center location cannot be displayed you can create a "Jogged" radius dimension.

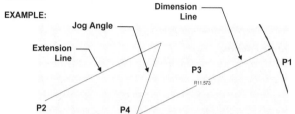

You can specify the jog angle in the Dimension Style / Symbols and Arrows tab.

1. Select the Jogged radius dimension using one of the following:

   Ribbon = Annotate / Dimension panel / 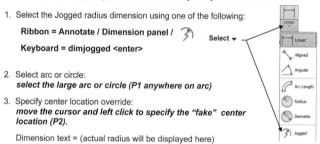 Select ▼

   Keyboard = dimjogged <enter>

2. Select arc or circle:
   select the large arc or circle (P1 anywhere on arc)

3. Specify center location override:
   move the cursor and left click to specify the "fake" center location (P2).

   Dimension text = (actual radius will be displayed here)

4. Specify dimension line location or [Mtext/Text/Angle]: *move the cursor and left click to specify the location for the dimension text (P3).*

5. Specify jog location: *move the cursor and left click to specify the location for the jog (P4).*

CONTROLLING THE JOG

You can set the Jog Angle and Height factor in:
Dimension Style / Symbols and Arrows tab

DIMENSIONING DIAMETERS

The **DIAMETER** dimensioning command should be used when dimensioning circles and arcs of <u>more than 180 degrees</u>. AutoCAD measures the selected circle or arc and displays the dimension text with the diameter symbol (Ø) in front of it.

1. Select the Diameter command using one of the following:

   Ribbon = Annotate tab / Dimension panel / 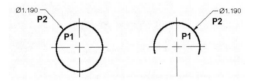 Select ▼

   Keyboard = Dimdiameter <enter>

2. Select arc or circle: *select the arc or circle (P1)* <u>*do not use object snap.*</u>
   Dimension text = *the diameter will be displayed here.*

3. Specify dimension line location or [Mtext/Text/Angle]: *place dimension text location (P2)*

Ø1.190
P2
P1

Ø1.190
P2
P1

<u>Center marks</u> are automatically drawn as you use the diameter dimensioning command.
If the circle already has a center mark or you do not want a center mark, set the center mark setting to **NONE** (Dimension Style / Symbols and Arrows tab) before using Diameter dimensioning.

DIMENSIONING DIAMETERS....continued

Controlling the diameter dimension appearance.

If you would like your Diameter dimensions to appear as shown in the two examples
below, you must change some setting in the Dimension Style **Fit tab**.

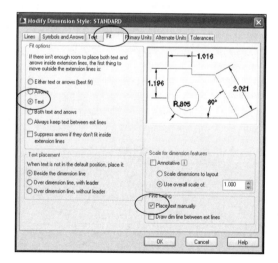

DIMENSIONING RADII

The **Radius** dimensioning command should be used when dimensioning arcs of <u>less than 180 degrees</u>. AutoCAD measures the selected arc and displays the dimension text with the radius symbol (R) in front of it.

1. Select the Radius command using one of the following:

   Ribbon = Annotate tab / Dimension panel / [icon] **Select ▼**

   Keyboard = Dimradius <enter>

2. Select arc or circle: **select the arc (P1) <u>do not use object snap.</u>**
   Dimension text = *the radius will be displayed here.*

3. Specify dimension line location or [Mtext/Text/Angle]: *place dimension text location (P2)*

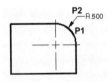

Center marks are automatically drawn as you use the radius dimensioning command.
If you do not want a center mark, set the center mark setting to **<u>NONE</u>**
(Dimension Style / Symbols and Arrows tab) before using Diameter dimensioning.

DIMENSIONING RADII

Controlling the radius dimension appearance.

If you would like your Radius dimensions to appear as shown in the example below, you must change the "**Fit Options**" and "**Fine Tuning**" in the **Fit tab** in your Dimension Style.

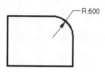

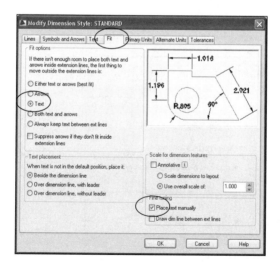

QUICK DIMENSION

Quick Dimension creates multiple dimensions with one command. Quick Dimension can create Continuous, Staggered, Baseline, Ordinate, Radius and Diameter dimensions. Qdim only works in modelspace. It is not trans-spatial.

The following is an example of how to use the "Continuous" option. Staggered, Baseline, Diameter and Radius are explained on the next page.

1. Select the **Quick Dimension** command using one of the following:

   Ribbon = Annotate tab / Dimension panel / ⟦⟧

   Keyboard = Qdim <enter>

   > Command: _qdim
   > Associative dimension priority = Endpoint
   > Select geometry to dimension

2. Select the objects to be dimensioned with a crossing window or pick each object

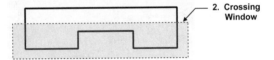

2. **Crossing Window**

3. Press <enter> to stop selecting objects.

4. Specify dimension line position, or
   [Continuous/Staggered/Baseline/Ordinate/Radius/Diameter/datumPoint/Edit/
   settings]<Continuous>: **Select "C" <enter> for Continuous.**

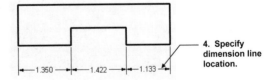

4. **Specify dimension line location.**

|← 1.350 →|← 1.422 →|← 1.133 →|

QUICK DIMENSION....continued

STAGGERED

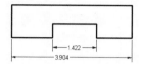

BASELINE

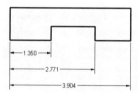

DIAMETER

1. You can use a crossing window. Qdim will automatically filter out any linear dims.
2. <u>Dimension Line Length </u>is determined by the "Baseline Spacing" setting in the Dimension Style.

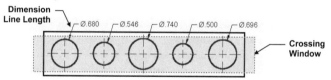

RADIUS

1. You can use a crossing window. Qdim will automatically filter out any linear dims.
2. Dimension Line Length is determined by the "Baseline Spacing" setting in the Dimension Style.

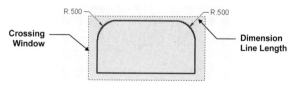

DIMENSION BREAKS

Occasionally extension lines overlap another extension line or even an object. If you do not like this you may use the **Dimbreak** command to break the intersecting lines. **Automatic** (described below) or **Manual** method may be used. You may use the **Restore** option to remove the break.

AUTOMATIC DIMENSION BREAKS

To create an automatically placed dimension break, you select a dimension and then use the Auto option of the DIMBREAK command.

Automatic dimension breaks are updated any time the dimension or intersecting objects are modified.

You control the size of automatically placed dimension breaks on the <u>Symbols and Arrows tab</u> of the Dimension Style dialog box.

The specified size is affected by the dimension break size, dimension scale, and current annotation scale for the current viewport.

1. Select the Dimbreak command using one of the following:

   Ribbon = Annotate tab / Dimension panel / [icon]

   Keyboard = Dimbreak <enter>

   Command: _Dimbreak
2. Select dimension to add / remove break or [Multiple]: *type M <enter>*

3. Select dimensions: *select dimension*

4. Select dimensions: *select another dimension*

5. Select dimensions: *select another dimension or <enter> to stop selecting*

6. Select object to break dimension or [Auto / Remove] <Auto>: *<enter>*

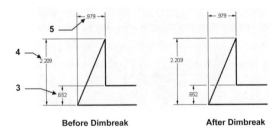

Before Dimbreak   **After Dimbreak**

DIMENSION BREAK....continued

MANUAL DIMENSION BREAK

You can place a dimension break by <u>picking two points</u> on the dimension, extension, or leader line to determine the size and placement of the break.

Dimension breaks that are added manually by picking two points are not automatically updated if the dimension or intersecting object is modified. So if a dimension with a manually added dimension break is moved or the intersecting object is modified, you might have to restore the dimension and then add the dimension break again.

The size of a dimension break that is created by picking two points is not affected by the current dimension scale or annotation scale value for the current viewport.

1. Select the Dimbreak command using one of the following:

    Ribbon = Annotate tab / Dimension panel / ⊥

    Keyboard = Dimbreak <enter>

Command: _Dimbreak
2. Select dimension to add / remove break or [Multiple]: *select the dimension*

3. Select object to break dimension or [Auto/Manual/Restore] <Auto>: *type M <enter>*

4. Specify first break point: *select the first break point location* (Osnap should be off)

5. Specify second break point: *select the second break point location*

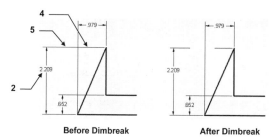

          Before Dimbreak   **After Dimbreak**

The following objects can be used as cutting edges when adding a dimension break:
Dimension, Leader, Line, Circle, Arc, Spline, Ellipse, Polyline, Text, and Multiline text.

DIMENSION BREAK....continued

REMOVE THE BREAK

Removing the break is easy using the <u>Restore</u> option.

1. Select the Dimbreak command using one of the following:

   Ribbon = Annotate tab / Dimension panel /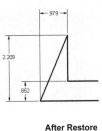

   Keyboard = Dimbreak <enter>

   Command: _DIMBREAK
2. Select dimension to add / remove break or [Multiple]: *type M <enter>*

3. Select dimensions: *select a dimension*

4. Select dimensions: *select a dimension*

5. Select dimensions: *select another dimension or <enter> to stop selecting*

6. Select object to break dimension or [Auto / Remove] <Auto>: *type R <enter>*

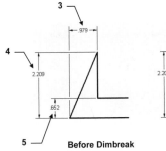

Before Dimbreak   **After Restore**

JOG A DIMENSION LINE

Jog lines can be added to linear dimensions. Jog lines are used to represent a
dimension value that does not display the actual measurement.

Before you add a jog, the Jog angle and the height factor of the jog should be set in
the <u>Symbols & Arrows tab</u> within the Dimension Style.

The height is calculated as a factor of the Text ht.

For example:
if the text height was .250 and the jog height factor
was 1.500, the jog would be .375
Formula: (.250 ht. X 1.500 jog ht. factor= .375).

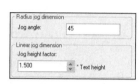

1. Select the Jogged Linear command

    Keyboard = Dimjogline <enter>

    Command: _DIMJOGLINE
2. Select dimension to add jog or [Remove]: **Select a dimension**

3. Specify jog location (or press ENTER): **Press <enter>**

4. <u>After you have added the jog you can position it by using grips and adjust the height
of the jog symbol using the Properties palette.</u>

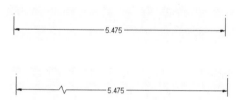

REMOVE A JOG
1. Select the Jogged Linear command
   Command: _DIMJOGLINE

2. Select dimension to add jog or [Remove]: **type R <enter>**

3. Select jog to remove: **select the dimension**

ADJUST DISTANCE BETWEEN DIMENSIONS

The Adjust Space command allows you to adjust the distance between existing parallel linear and angular dimensions, so they are equally spaced. You may also align the dimensions to create a string.

1. Select the Dimension Space command using one of the following:

   Ribbon = Annotate tab / Dimension panel /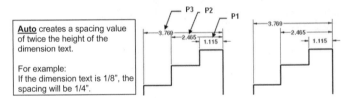

   Keyboard = Dimspace <enter>

   Command: _DIMSPACE
2. Select base dimension: *Select the dimension that you want to use as the base dimension when equally spacing dimensions. (P1)*
3. Select dimensions to space: *select the next dimension to be spaced. (P2)*
4. Select dimensions to space: *select the next dimension to be spaced. (P3)*
5. Select dimensions to space: *continue selecting or press <enter> to stop*
6. Enter value or [Auto] <Auto>: *enter a value or press <enter> for auto*

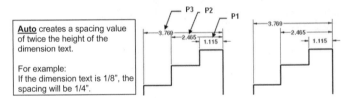

> **Auto** creates a spacing value of twice the height of the dimension text.
>
> For example:
> If the dimension text is 1/8", the spacing will be 1/4".

Aligning dimensions

Follow the steps shown above but when asked for the value enter "0" <enter>.

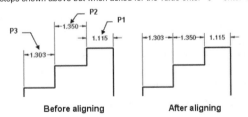

Before aligning   After aligning

MULTILEADER

A multileader is a single object consisting of a **pointer**, **line**, **landing** and **content**.
You may draw a multileader, Pointer first, tail first or content first.

To select the Multileader use:
Annotate tab / Leaders panel

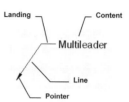

Create a single Multileader

1. Select the **Multileader** tool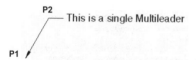

Command: _mleader
2. Specify leader arrowhead location or [leader Landing first/Content first/Options]
                                       <Options>: *specify arrowhead location (P1)*
3. Specify leader landing location: *specify the landing location (P2)*
4. The Multitext editor appears. *Enter text then select "Close Text Editor"*

P2
— This is a single Multileader

P1

Add leader lines to existing Multileader

1. Select the **Add Leader** tool
2. Select a multileader: *click on the existing multileader line (P1)*
3. Specify leader arrowhead location: *specify arrowhead location (P2)*
4. Specify leader arrowhead location: *specify next arrowhead location (P3)*
5. Specify leader arrowhead location: *specify next arrowhead location (P4)*
6. Specify leader arrowhead location: *press <enter> to stop*

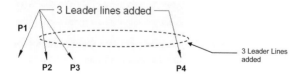

MULTILEADER....continued

Remove a leader line from an existing Multileader

1. Select the **Remove Leader** tool

2. Select a multileader: *click anywhere on the existing multileader*
3. Specify leaders to remove: *select the leader line to remove*
4. Specify leaders to remove: *press <enter> to stop*

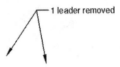

Align Multileaders

1. Select the **Align Multileader** tool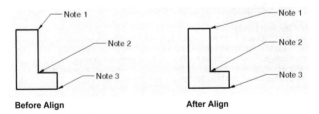

Command: _mleaderalign
2. Select multileaders: *select the leaders to align (Note 1 and Note 3)*
3. Select multileaders: *press <enter> to stop selecting*
   Current mode: Use current spacing
4. Select multileader to align to or [Options]: *select the leader to align to (Note 2)*
5. Specify direction: *move the cursor and press left mouse button*

Note: **Collect Multileader** works best with Blocks and will be discussed in Section 4.

CREATE A MULTILEADER STYLE

You may create **Multileader styles** to control it's appearance.
This will be similar to creating a dimension style.

1. Open a drawing.

2. Select the **Multileader Style Manager** by selecting the ↘ on the Leader panel.

3. Select the **New** button.
4. Enter the New Style name: **Class ML style**
5. Select the **Continue** button.

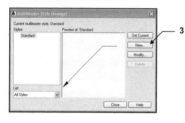

6. Select the **Leader Format** tab and change the settings as shown below.

<u>General</u> – You may set the leader
line to straight or spline (curved).
You may change the color, linetype
and lineweight.

<u>Arrowhead</u> – Select the symbol for
the pointer and size of pointer.

<u>Leader break</u> - specify the size of
the gap in the leader line if
Dimension break is used.

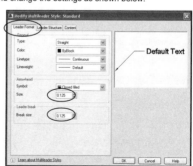

CREATE A MULTILEADER STYLE....continued

7. Select the **Leader Structure** tab and change the settings as shown below.

Constraints – Specify how many line segments to allow and on what angle.

Landing Settings – Specify the length of the Landing (the horizontal line next to the note) or turn it off by unchecking the "Automatically include landing" box.

Scale – "Annotation" will be discussed in Section 10.

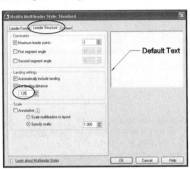

8. Select the **Content** tab and change the settings as shown below.

Multileader type – Select what to attach to the landing.

Text Options – Specify the leader note appearance.

Leader Connection – Specify how the note attaches to the landing.

9. Select the **OK** button.

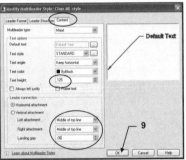

The **Class ML style** should be listed in the "Style" area.

10. Select **Set Current** button.

11. Select **Close** button.

12. **Save** your drawing again.
Now whenever you use this drawing the
Multileader style **Class ML style** will be there.

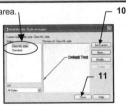

IGNORING HATCH OBJECTS

Occasionally, when you are dimensioning an object that has "Hatch Lines", your cursor
will snap to the Hatch Line instead of the object that you want to dimension.
To prevent this from occurring, select the option **"IGNORE HATCH OBJECTS"**.

EXAMPLE:

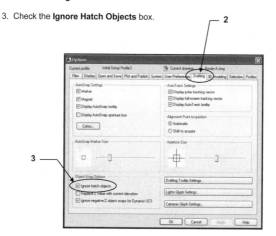

"Ignore Hatch Objects"
option <u>not</u> <u>selected</u>

1.414

"Ignore Hatch Objects"
option <u>selected</u>

1.679

How to select the "IGNORE HATCH OBJECTS" option

1. Select **Browser Menu / Options**

2. Select **Drafting** tab.

3. Check the **Ignore Hatch Objects** box.

2

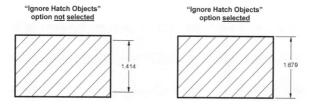

3

Notes

Section 4
Drawing Entities

ARC

There are 10 ways to draw an ARC in AutoCAD. Not all of the ARCS options are easy to create so you may find it is often easier to **trim a Circle** or use the **Fillet** command.

On the job, you will probably only use 2 of these methods. Which 2 depends on the application.

An **ARC** is a <u>segment</u> of a circle and <u>must be less than 360 degrees</u>.

Most ARCS are drawn counter-clockwise but some may be drawn clockwise by entering a negative input.

1. Select the Arc Command using one of the following:

   Ribbon = Home tab / Draw panel / [arc icon ▼]

   Keyboard = A <enter>

2. Refer to the following pages for examples of each method listed below.

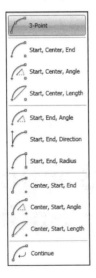

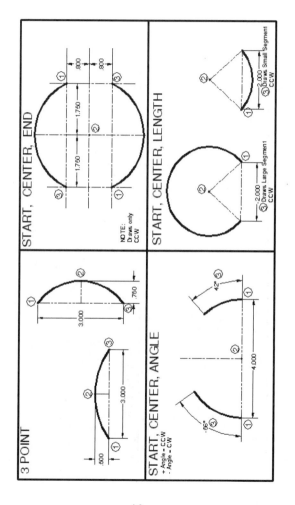

START, CENTER, END

NOTE:
Draws only
CCW

START, CENTER, LENGTH

③ Draws Small Segment
CCW

② Draws Large Segment
CCW

3 POINT

START, CENTER, ANGLE

+ Angle - CCW
- Angle - CW

4-3

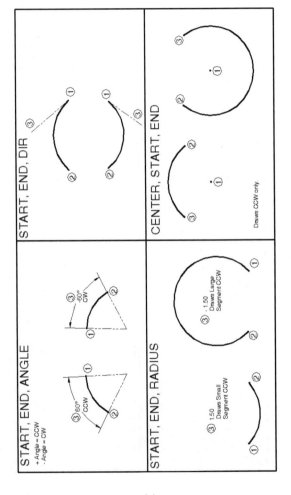

START, END, DIR

START, END, ANGLE

+ Angle = CCW
- Angle = CW

③ 60°
CCW

③ -60°
CW

CENTER, START, END

Draws CCW only.

START, END, RADIUS

③ 1.50
Draws Small
Segment CCW

③ -1.50
Draws Large
Segment CCW

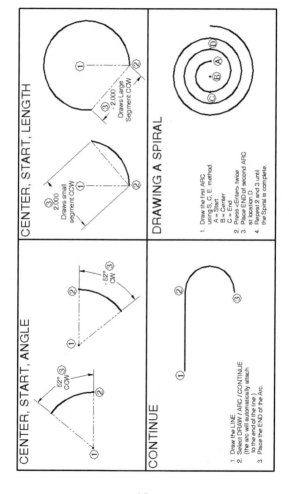

CENTER, START, ANGLE

52° CCW ③

- 52° CW ③

CENTER, START, LENGTH

③ 2.000
Draws small
segment CCW ①

- 2.000
Draws Large
Segment CCW

CONTINUE

1. Draw the LINE
2. Select DRAW / ARC / CONTINUE
   (the arc will automatically attach
   to the end of the line)
3. Place the END of the Arc.

DRAWING A SPIRAL

1. Draw the first ARC
   using S, C, E. method.
   A = Start
   B = Center
   C = End
2. Press <Enter> twice
3. Place END of second ARC
   at location D.
4. Repeat 2 and 3 until
   the Spiral is complete.

4-5

BLOCKS

A **BLOCK** is a group of objects that have been converted into ONE object. A Symbol, such as a transistor, bathroom fixture, window, screw or tree, is a typical application for the block command. First a BLOCK must be created. Then it can be INSERTED into the drawing. An inserted Block uses less file space than a set of objects copied.

CREATING A BLOCK

1. First draw the objects that will be converted into a Block.

   For this example a circle and 2 lines are drawn.

2. Select the **BLOCK MAKE** command using one of the following:

   Ribbon = Insert tab / Block panel /

   Keyboard = B <enter>

3. Enter the New Block name in the **Name** box.

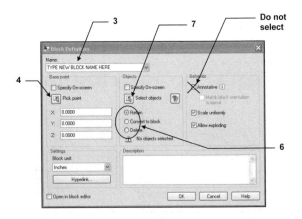

BLOCKS....continued

4. Select the **Pick Point** button. (Or you may type the X, Y and Z coordinates.)
   The Block Definition box will disappear and you will return temporarily to the drawing.

5. Select the location where you would like the insertion point for the Block.
   Later when you insert this block, the block will appear on the screen attached to the cursor at this insertion point. Usually this point is the CENTER, MIDPOINT or ENDPOINT of an object.

Notice the coordinates for the base point are now displayed. (Don't worry about this. Use Pick Point and you will be fine)

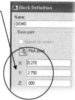

6. Select an option.

   It is important that you select one and understand the options below.

   Retain
   If this option is selected, the original objects will stay visible on the screen after the block has been created.

   Convert to block
   If this option is selected, the original objects will disappear after the block has been created, but will immediately reappear as a block. It happens so fast you won't even notice the original objects disappeared.

   Delete
   If this option is selected, the original objects will disappear from the screen after the block has been created. (This is the one I use most of the time)

7. Select the **Select Objects** button.

   The Block Definition box will disappear and you will return temporarily to the drawing.

8. Select the objects you want in the block, then press <enter>.

Selection Window

BLOCKS....continued

The Block Definition box will reappear and the objects you selected should be
illustrated in the Preview Icon area.

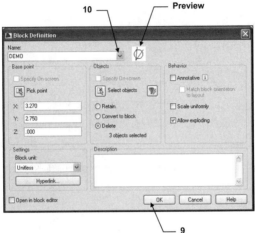

9. Select the **OK** button.
   The new block is now stored in the drawing's block definition table.

10. To verify the creation of this Block, select [⟐] **Create** again, and select the
    Name (▼). A list of all the blocks, in this drawing, will appear.

BLOCKS....continued

ADDITIONAL DEFINITIONS OF OPTIONS

Block Units
You may define the units of measurement for the block. This option is used with the "Design Center" to drag and drop with Autoscaling. The Design Center is an advanced option and is not discussed in this book.

Hyperlink
Opens the **insert Hyperlink dialog box** which you can use to associate a hyperlink with the block.

Description
You may enter a text description of the block.

Scale Uniformly
Specifies whether or not the block is prevented from being scaled non-uniformly during insertion.

Allow Exploding
Specifies whether or not the block can be exploded after insertion.

HOW LAYERS AFFECT BLOCKS

If a block is created on Layer 0:

1. When the block is inserted, it will take on the properties of the current layer.
2. The inserted block will reside on the layer that was current at the time of insertion.
3. If you Freeze or turn Off the layer the block was inserted onto, the block will disappear.
4. If the Block is **Exploded**, the objects included in the block will revert to their original properties of layer 0.

If a block is created on Specific layers:

1. When the block is inserted, it will retain its own properties. It **will not** take on the properties of the current layer.
2. The inserted block **will reside** on the current layer at the time of insertion.
3. If you **freeze** the layer that was current at the time of insertion the block will disappear.
4. If you turn **off** the layer that was current at the time of insertion the block will not disappear.
5. If you **freeze** or turn **off** the blocks original layers the block will disappear.
6. If the Block is **Exploded**, the objects included in the block will go back to their original layer.

INSERTING BLOCKS

A **BLOCK** can be inserted at any location within the drawing. When inserting a Block you can **SCALE** or **ROTATE** it.

1. Select the **INSERT** command using one of the following:

   Ribbon = Insert tab / Block panel /

   Keyboard = Insert <enter>

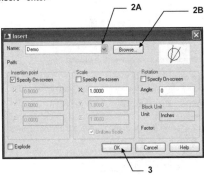

2. Select the **BLOCK** name.
   a. If the block is already in the drawing that is open on the screen, you may select the block from the drop down list shown above
   b. If you want to insert an entire drawing, select the Browse button to locate the drawing file.

3. Select the **OK** button.

   This returns you to the drawing and the selected block should be attached to the cursor.

4. Select the insertion location for the block by moving the cursor and pressing the left mouse button or typing coordinates.

   Command: _insert
   Specify insertion point or **[Basepoint/Scale/X/ Y/Z/Rotate]:**

   NOTE: If you want to change the **basepoint, scale** or **rotate** the block before you actually place the block, press the right hand mouse button and you may select an option from the menu or select an option from the command line menu shown above.

   You may also "**preset**" the insertion point, scale or rotation. This is discussed on the next page.

INSERTING BLOCKS....continued

PRESETTING THE <u>INSERTION POINT</u>, <u>SCALE</u> or <u>ROTATION</u>

You may preset the **Insertion point, Scale or Rotation** in the **INSERT** box instead of at the command line.

1. Remove the check mark from any of the **"Specify On-screen"** boxes.
2. Fill in the appropriate information describe below:

Insertion point
Type the X and Y coordinates <u>from the Origin</u>. The Z is for 3D only.
The example below indicates the block's insertion location will be 5 inches in the X direction and 3 inches in the Y direction, <u>from the Origin</u>.

Scale
You may scale the block proportionately by typing the scale factor in the X box and then check the <u>Uniform Scale box</u>. If you selected "Scale uniformly" box when creating the block this option is unnecessary and not available.
If the block is to be scaled non-proportionately, type the different scale factors in both X and Y boxes.
The example below indicates that the block will be scale proportionate at a factor of 2.

Rotation
Type the desired rotation angle relative to its current rotation angle.
The example below indicates the block will be rotated 45 degrees from its originally created angle orientation.

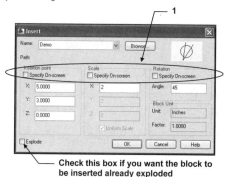

Check this box if you want the block to be inserted already exploded

RE-DEFINING A BLOCK

How to change the design of a block previously inserted.

1. Select **Block Editor** using one of the following:

   Ribbon = Insert tab / Block panel /
   Keyboard = bedit

   The "Edit Block Definition" dialog box will appear.

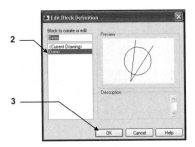

2. Select the name of the Block that you wish to change.

3. Select the **OK** button.

4. The Block that you selected should appear large on the screen.
   You may now make any additions or changes to the block. You can change tabs and use other panels such as Draw and Modify. But you must return to the **Block Editor** tab to complete the process.

5. Return to the **Block Editor** button if you selected any other tab.

6. Select the **Save block** tool from the **Open / Save** panel.

7. Select the **Close** tool.

You will be returned to the drawing and **ALL** of the blocks with the same name will be updated with the changes that you made.

PURGE UNWANTED and UNUSED BLOCKS

You can remove a block reference from your drawing by erasing it; however, the block definition remains in the drawing's block definition table. To remove any **unused** blocks, dimension styles, text styles, layers and linetypes you may use the **Purge** command.

How to delete unwanted and unused blocks from the current drawing.

1. Select the **Purge** command using one of the following:

   Ribbon = None

   Application Menu = Drawing Utilities / Purge

   Keyboard = Purge <enter>

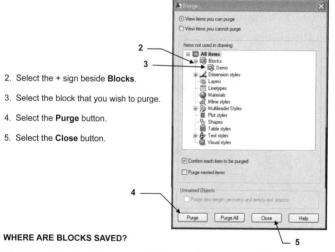

2. Select the + sign beside **Blocks**.

3. Select the block that you wish to purge.

4. Select the **Purge** button.

5. Select the **Close** button.

WHERE ARE BLOCKS SAVED?

When you create a Block it is saved **within the drawing you created it in**.
(If you open another drawing you will not find that block.)

MULTILEADER and BLOCKS

In Section 3 you learned about Multileaders and how easy and helpful they are to use. Now you will learn another user option within the Multileader Style Manager that allows you to <u>attach a **pre-designed Block**</u> to the landing end of the Leader. To accomplish this, you must <u>first create the style</u> and <u>then you may use it</u>.

Here are a few examples of multileaders with pre-designed blocks attached:

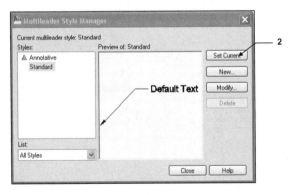

STEP 1. CREATE A NEW MULTILEADER STYLE

1. Select the **Multileader Style Manager** tool using

    Ribbon = Annotate tab / Leaders panel / ↘

2. Select the **New** button.

MULTILEADER and BLOCKS....continued

3. Enter **New Style Name**

4. Select a style to **Start with**:

5. Select **Annotative** box

6. Select **Continue** button.

7. Select the **Content** tab.

8. **Multileader Type**: Select **Block** from drop down menu.

9. **Source Block**: Select **Circle** from the drop down menu.
   Note: you have many choices here. These are AutoCAD pre-designed blocks <u>with</u>
   <u>Attributes.</u>

10. **Attachment**: Select **Center Extents**.
    Note: this selection works best with Circle but you will be given <u>different choices</u>
    depending on which <u>Source block you select.</u>

11. **Color**: Select **Bylayer**
    Bylayer works best. It means, it acquires the color of the current layer setting.

12. **Scale:** Select 1.000

13. Select **OK** button.

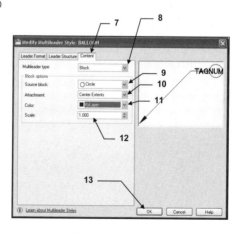

MULTILEADER and BLOCKS....continued

Your new multileader style should be displayed in the Styles list.

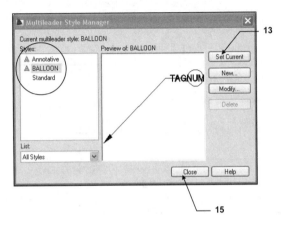

14. Select **Set Current**

15. Select **Close** button.

MULTILEADER and BLOCKS....continued

STEP 2. USING THE MULITLEADER WITH A BLOCK STYLE.

1. Select Layer **Symbol** from the Layer manager.

2. Select the **Style** from the Multileader drop down list

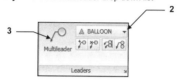

3. Select the **Multileader** tool.

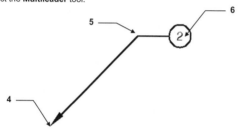

4. Specify leader arrowhead location or [leader Landing first/Content first/Options] <Options>: *place the location of the arrowhead*

5. Specify leader landing location: *place the location of the landing*

> *The next step is where the pre-assigned "Attributes" activate.*
> *Refer to the Advanced Workbook for Attributes.*

6. Enter attribute values
   Enter tag number <TAGNUMBER>: *type number or letter <enter>*

COLLECT MULTILEADER

In Section 3 you learned how to ADD, REMOVE and ALIGN multileaders. Now you will learn how to use the **COLLECT** multileader tool.

If you have multiple Leaders pointing to the same location or object you may wish to COLLECT them into one Leader.

Example:

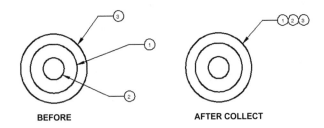

BEFORE   AFTER COLLECT

HOW TO USE COLLECT MULTILEADER

1. Select the Collect Mutileader tool

2. Select the Multileaders that you wish to combine.
   Note: Select them one at a time <u>in the order you wish them to display</u>.
       Such as (1, 2, 3, A, B, C, etc)

3. Place the combined Leader location. (**Ortho** should be **OFF**)

CENTERMARK

CENTERMARKS can ONLY be drawn with circular objects like Circles and Arcs.
You set the size and type.

The Center Mark has three <u>types</u>, **None, Mark** and **Line** as shown below.

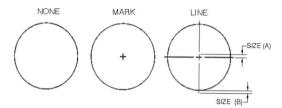

What does "SIZE" mean?

The **size** setting determines both, (A) the length of half of the intersection line and
(B) the length extending beyond the circle. (See above right)

Where do you set the CENTERMARK "TYPE" and "SIZE"

1. Select the **Dimension Style** command.
2. Select: **Modify or Override.**
3. Select: **SYMBOLS and ARROWS** tab
4. Select the **Center mark type**
5. Set the **Size**.

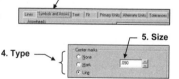

To draw a CENTER MARK

1. Select the **Center mark** command using one of the following:

   Ribbon = Annotation tab / Dimension ▾ panel / ⊕

   Keyboard = DCE <enter>

2. Select arc or circle: *select the arc or circle with the cursor.*

CIRCLE

There are 6 options to create a circle.

The default option is **"Center, radius"**. (Probably because that is the most common method of creating a circle.)

We will try the **"Center, radius"** option first.

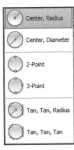

1. Start the **Circle** command by using one of the following:

   Ribbon = Home tab / Draw panel /

   Keyboard = C <enter>

2. The following will appear on the command line:
   Command: _circle Specify center point for circle or [3P/2P/Ttr (tan tan radius)]:

3. Locate the center point for the circle by moving the cursor to the desired location in the drawing area **(P1)** and press the left mouse button.

4. Now move the cursor away from the center point and you should see a circle forming.

5. When the circle is the size desired **(P2)**, press the left mouse button, or type the radius and then press <enter>.

Note: To use one of the other methods described below, first select the Circle command, then select one of the other Circle options.

Center, Radius: (Default option)
1. Specify the center (P1) location.

2. Specify the Radius (P2).

   (Define the Radius by moving the cursor or typing radius) P2

Center, Diameter:
1. Specify the center (P1) location.

2. Specify the Diameter (P2). (Define the Diameter by moving the cursor or typing Diameter)

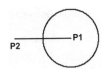

CIRCLE....continued

2 Points:
1. Select the 2 point option
2. Specify the 2 points (P1 and P2) that will determine the Diameter .

3 Points:
1. Select the 3 Point option
2. Specify the 3 points (P1, P2 and P3) on the circumference.
   The Circle will pass through all three point

Tangent, Tangent, Radius:
1. Select the Tangent, Tangent, Radius option .
2. Select two objects (P1 and P2) for the Circle to be tangent to by placing the cursor on the object and pressing the left mouse button
3. Specify the radius.

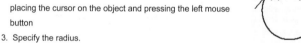

Tangent, Tangent, Tangent:
1. Select the Tangent, Tangent, Tangent option
2. Specify three objects (P1, P2 and P3) for the Circle to be tangent to by placing the cursor on each of the objects and pressing the left mouse button. (AutoCAD will calculate the diameter.)

DONUT

A Donut is a circle with **width**. You will define the **Inside** and **Outside** diameters.

1. Select the **DONUT** command using one of the following:

   Ribbon = Home tab / Draw panel /

   Keyboard = DO <enter>

2. The following prompts will appear on the command line:

   Command: _donut
   Specify inside diameter of donut: *type the inside diameter <enter>*
   Specify outside diameter of donut: *type the outside diameter <enter>*
   Specify center of donut or <exit>: *place the center of the first donut*
   Specify center of donut or <exit>: *place the center of the second donut or*
   <enter> to stop

Controlling the "FILL MODE"

1. Command: *type FILL <enter>*

2. Enter mode [ON / OFF] <OFF>: *type ON or OFF <enter>*

3. Select type *REGEN <enter>* to regenerate the drawing to show the latest setting of the *FILL* mode.

FILL = ON

FILL = OFF

ELLIPSE

There are 3 methods to draw an Ellipse. You may (1) specify 3 points of the axes, (2) define the center point and the axis points or (3) define an elliptical Arc.
The following 3 pages illustrates each of the methods.

AXIS END METHOD

1. Select the **ELLIPSE** command using one of the following:

Ribbon = Home tab / Draw panel / Axis, End

Keyboard = EL <enter>

2. The following prompts will appear on the command line:

Command: _ellipse
Specify axis endpoint of ellipse or [Arc/Center]: *place the first point of either the major or minor axis (P1).*
Specify other endpoint of axis: *place the other point of the first axis (P2)*
Specify distance to other axis or [Rotation]: *place the point perpendicular to the first axis (P3).*

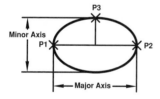

Specifying <u>Major Axis first </u>(P1/P2),
then Minor Axis (P3)

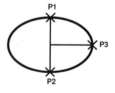

Specifying <u>Minor Axis first </u>(P1/P2),
then Major Axis (P3)

Continued on the next page...

ELLIPSE....continued

CENTER METHOD

1. Select the **ELLIPSE** command using one of the following:

   Ribbon = Home tab / Draw panel /

   Keyboard = EL <enter> C <enter>

2. The following prompts will appear on the command line:

   Command: _ellipse
   Specify axis endpoint of ellipse or [Arc/Center]: _c
   Specify center of ellipse: *place center of ellipse (P1)*
   Specify endpoint of axis: *place first axis endpoint (either axis) (P2)*
   Specify distance to other axis or [Rotation]: *place the point perpendicular to the
   first axis (P3)*

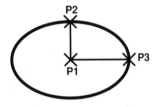

Continued on the next page...

ELLIPSE....continued

ELLIPTICAL ARC METHOD

1. Select the **ELLIPSE** command using one of the following:

   Ribbon = Home tab / Draw panel /  Elliptical Arc

   Keyboard = EL <enter> A <enter>

2. The following prompts will appear on the command line:
   Command: _ellipse
   Specify axis endpoint of ellipse or [Arc/Center]: _a
   Specify axis endpoint of elliptical arc or [center]: *type C <enter>*
   Specify center of axis: *place the center of the elliptical arc (P1)*
   Specify endpoint of axis: *place first axis point (P2)*
   Specify distance to other axis or [Rotation]: *place the endpoint perpendicular to the first axis (P3)*

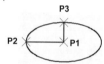

   Specify start angle or [Parameter]: *place the start angle (P4)*

   Specify end angle or [Parameter/Included angle]: *place end angle (P5)*

GRADIENT FILL

Note: Gradient is not available in the LT version.

Gradients are fills that gradually change from dark to light or from one color to another. Gradient fills can be used to enhance presentation drawings, giving the appearance of light reflecting on an object, or creating interesting backgrounds for illustrations.

Gradients are definitely fun to experiment with but you will have to practice to achieve complete control.

1. Select the HATCH command using one of the following:

   Ribbon = Home tab / Draw panel / ▨

   Keyboard = BH <enter>

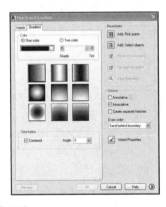

2. Select the Gradient tab

3. Select **Boundary** using either <u>Add: Pick Points</u> or <u>Select Objects</u>.

4. Choose the Gradient settings described on the next page.

5. Preview

6. Accept or make changes.

GRADIENT FILL....continued

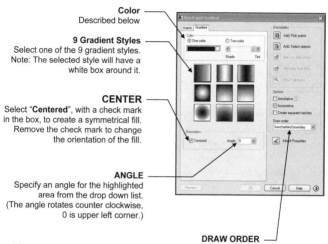

Color
Described below

9 Gradient Styles
Select one of the 9 gradient styles.
Note: The selected style will have a
white box around it.

CENTER
Select "**Centered**", with a check mark
in the box, to create a symmetrical fill.
Remove the check mark to change
the orientation of the fill.

ANGLE
Specify an angle for the highlighted
area from the drop down list.
(The angle rotates counter clockwise,
0 is upper left corner.)

DRAW ORDER
When you create a gradient, it appears in front of the boundary and
sometimes obscures the object's outline. To send the gradient hatch
behind the boundary, specify the Draw Order.

ONE COLOR
Click the (...) button to the right of the color swatch to
open the Select Color dialog box.
Choose the color you want.

Use the "**Shade and Tint**" slider to choose
the gradient range from lighter to darker.

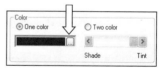

TWO COLOR
Click each (...) button to choose a color.
When you choose two colors the transition is
both from light to dark and from the first color
to the second.

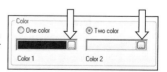

HATCH

The **HATCH** command is used to create hatch lines for section views or filling areas with specific patterns.

To draw **hatch** you must start with a closed boundary. A closed boundary is an area completely enclosed by objects. A rectangle would be a closed boundary. You simply pick inside the closed boundary. HATCH locates the area and automatically creates a temporary polyline around the outline of the hatch area. After the hatch lines are drawn in the area, the temporary polyline is automatically deleted. (Polylines are discussed in a future lesson)

Note: A Hatch set is one object. If you explode it, it will return to multiple objects such as lines.

It is good drawing management to always place gradient fill on it's own layer.
Use Layer Hatch

You may also make gradient fill appear or disappear with the **FILL** command

1. Select the Hatch command using one of the following:

   Ribbon = Home tab / Draw panel /

   Keyboard = BH or hatch <enter>

2. Select the hatch **Type**

   (**Predefined** and **User defined** described on the following pages)

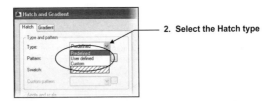

2. Select the Hatch type

HATCH….continued

PREDEFINED

AutoCAD has many predefined hatch patterns.
These patterns are stored in the acad.pat and acadiso.pat files.
(You may also purchase patterns from other software companies.)

> **Note:** Using Hatch patterns will greatly increase the size of the drawing file.
> So use them conservatively.

To select a pattern by name, click on the "Pattern" down arrow. A drop-down list of available patterns will appear.

To select a pattern by appearance, click on the (…) button. This will display the Hatch Pattern Palette" dialog box.

This box displays the name of the Pattern you selected.

This determines the rotation angle of the pattern. A pre-designed pattern has a default angle of 0. If you change this Angle it will rotate the pattern relative to its original design.

The selected pattern is displayed here.

The value in this box is the scale factor. This will be discussed more later.

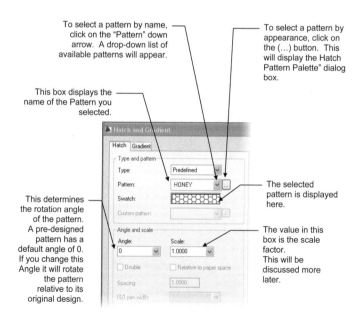

4-29

HATCH....continued

USER DEFINED

This selection allows you to simply draw continuous lines. (No special pattern)
You specify the **Angle** and the **Spacing** between the lines.

(This selection does not increase the size of the drawing file like Predefined)

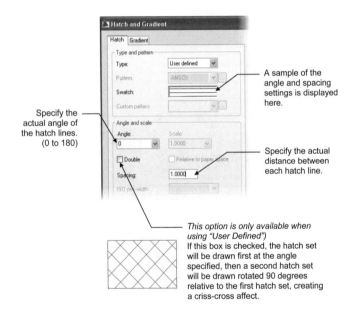

A sample of the angle and spacing settings is displayed here.

Specify the actual angle of the hatch lines. (0 to 180)

Specify the actual distance between each hatch line.

This option is only available when using "User Defined")
If this box is checked, the hatch set will be drawn first at the angle specified, then a second hatch set will be drawn rotated 90 degrees relative to the first hatch set, creating a criss-cross affect.

HATCH....continued

3. Select the **Options**

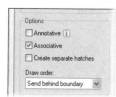

Annotative: Automatically adjusts the scale to match the current Annotative scale.
This option will be discussed in Section 10.

Associative: The hatch set is one entity and if the boundary size is changed the hatch will automatically change to the new boundary shape.

Create separate hatches: Controls whether HATCH creates a single hatch object or separate hatch objects when selecting several closed boundaries.

4. Select the Area that you would like to Hatch.
   This operation is called "Selecting the boundary".

Add: Pick Points:
Select the **Pick Points** box then select a point inside the area you want to hatch. A boundary will automatically be determined.

Add: Select Objects:
Select the boundary by selecting the object(s).

Example of Pick Points:

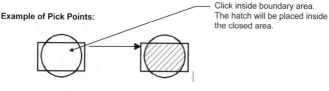

Click inside boundary area.
The hatch will be placed inside the closed area.

Example of Select Objects:

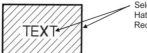

Select the Text and Rectangle
Hatch will be placed inside the Rectangle but outside the text.

HATCH....continued

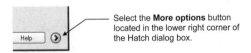

Select the **More options** button located in the lower right corner of the Hatch dialog box.

5. Boundary Retention
   It is important to select **Retain** or **Not retain** the boundary of the hatch.
   The hatch boundary can be retained, or disappear, if you erase the hatch lines.

6. If the area you selected to hatch is not completely closed (gaps) AutoCAD attempts
   to show you where the problem may have occurred. **Red circles** appear around
   endpoints near the gap and a warning appears.

If a warning appears you may cancel the hatch command and close the gap or
Increase the Gap Tolerance. The Gap tolerance can be set to a value from 0 to 5000.
Any gaps equal to or smaller than the value you specify are ignored and the boundary
is treated as closed.

HATCH....continued

7. Preview the Hatch

    a. After you have selected the boundary, press the right mouse button and select
       "**Preview**" from the short cut menu. This option allows you to preview the hatch
       set before it is actually applied to the drawing.

    b. If the preview is not what you expected, press the **ESC** key, make the changes
       and preview again.

    c. When you are satisfied, **right click** or press **<enter>** to accept.

       Note: It is always a good idea to take the extra time to preview the hatch.
              It will actually save you time in the long run.

HATCH....continued

HATCH - SOLID

If you would like to fill an area with a solid fill you should use **Predefined - Solid**.

> Note: The **Solid** pattern is a little difficult to find because it is not in
> alphabetical order. You will find it at the top of the pattern list.

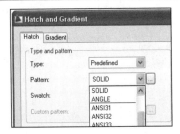

You apply **Solid** hatch just like you would any other hatch pattern.
Select the pattern, select the boundary, preview and accept.

Examples of Solid hatch pattern:

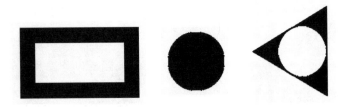

EDITING HATCH

HATCHEDIT allows you to edit an existing hatch pattern in the drawing.
You simply double click on the hatch set that you want to change or type
he <enter> and select the hatch set you want to change and the hatch dialog box will
appear. Make the changes, preview and accept.

CHANGING THE BOUNDARY
If you change the shape of the boundary the hatch will conform to the new shape
if the hatch is ASSOCIATIVE. This means the hatch is associated to the object.

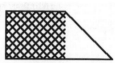

Original Boundary   **New Boundary**   **New Boundary**
with Associative hatch   **with non-Associative hatch**

TRIMMING HATCH
You may trim a hatch set just like any other object.

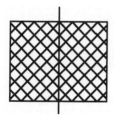
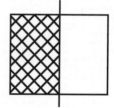

DRAWING LINES

A **Line** can be **one segment** or a **series of connected segments.**
But each segment is an individual object.

One Segment   **Series of connected Segments**

Start the Line command using one of the following methods:

Ribbon = Home tab / Draw Panel /

Keyboard = L <enter>

Lines are drawn by specifying the locations for each endpoint.
Move the cursor to the location of the **"first"** endpoint (1) then press the left mouse
button. Move the cursor again to the **"next"** endpoint (2) and press the left mouse
button. Continue locating **"next"** endpoints until you want to stop.

There are 2 ways to **Stop drawing a line**:
1. Press <enter> key.
2. Press <Space Bar>

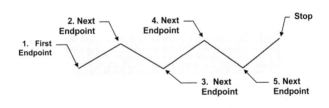

```
Helpful hint:
To quickly repeat the Line command, press the Spacebar.
```

Continued on the next page...

DRAWING LINES....continued

Horizontal and Vertical Lines

To draw a Line perfectly Horizontal or Vertical select the **Ortho** mode by selecting the **Ortho button** on the Status Bar or pressing the **F8 key**.

  or

Try the following exercise:

1. Select the Line command.

2. Place the **First endpoint**

3. **Turn Ortho ON** by selecting the Ortho button or F8. (The "Ortho" button will change to blue when ON.)

4. Move the cursor to the right and press the left mouse button to place the **next endpoint.** (The line should appear perfectly horizontal.)

5. Move the cursor down and press the left mouse button to place the next endpoint. (The line should appear perfectly vertical)

6. Now turn <u>Ortho **OFF**</u> by selecting the Ortho button. (The "Ortho" button will change to gray when OFF.)

7. Now move the cursor up and to the right at an upper ward angle. (The line should move freely now)

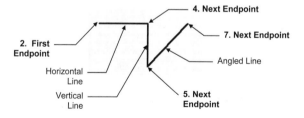

Ortho can be turned ON or OFF at any time while you are drawing. It can also be turned OFF temporarily by holding down the **Shift** key. Release the Shift key to resume.

Continued on the next page...

DRAWING LINES....continued

Closing Lines

If you have drawn 2 or more line segments, the <u>endpoint of the last line segment</u> can
be connected automatically to the <u>first endpoint</u> using the **Close** option.

Try the following exercise:

1. Select the Line command.

2. Place the **First endpoint**

3. Place the next endpoint

4. Place the next endpoint

5. Type **C** <enter>

   Or

5. <u>Press the right mouse button</u> and select **Close** from the **Shortcut** menu.

What is the Shortcut Menu?
The <u>Shortcut menu</u> gives you quick access to command options.

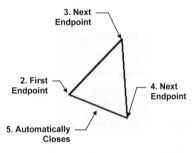

If you have Dynamic Input <u>OFF</u>:
Press the right mouse button.
The shortcut menu will appear.
Select an option.

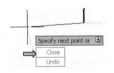

If you have Dynamic Input <u>ON</u>:
You may use the right mouse
button or press the down arrow ↓
and the options will appear below
the Dynamic Input prompt.

POINT

Points are used to locate a point of reference or location. A **Point** may be represented by one of many **Point Styles** shown below in the <u>Point Style box</u>.

The only object snap option that can be used with Point is **Node**.

HOW TO <u>USE</u> THE POINT COMMAND

1. Select the **POINT** command using one of the following:

    Ribbon = Home tab / Draw panel ▾/ ⟨ · ⟩

    Keyboard = PO <enter>

2. The following prompts will appear on the command line:

    Command: _point
    Current point modes: PDMODE=3 PDSIZE=0.000
    Specify a point: *place the point location*
    Specify a point: *place another point or press the "ESC" key to stop*

HOW TO <u>SELECT</u> A "POINT STYLE"

1. Open the Point Style dialog box:

    Keyboard = ddptype <enter>

2. The Point Style dialog box will appear.

3. Select a point style tile.

4. Select the **OK** button.

Point Size:
Set Size Relative to Screen
Sets the point display size as a percentage of the screen size. The point display does not change when you zoom in or out

Set Size in Absolute Units
Sets the point display size as the actual units you specify under Point Size. Points are displayed larger or smaller when you zoom in or out.

POLYGON

A polygon is an object with multiple edges (flat sides) of equal length. You may specify from 3 to 1024 sides. A polygon appears to be multiple lines but in fact it is one object. You can specify the <u>center and a radius</u> or the <u>edge length</u>. The <u>radius</u> size can be specified <u>Inscribed)</u> or <u>Circumscribed</u>.

CENTER, RADIUS METHOD
1. Select the **Polygon** command using one of the following:

   Ribbon = Home tab / Draw panel ▾ /

   Keyboard = POL <enter>

2. The following prompts will appear on the command line:

   _polygon Enter number of sides <4>: **type number of sides <enter>**
   Specify center of polygon or [Edge]: **specify the center location (P1)**
   Enter an option [Inscribed in circle/Circumscribed about circle]<I>:**type I or C<enter>**
   Specify radius of circle: **type radius or locate with cursor. (P2)**

Note:
The dashed circle is shown only as a reference to help you visualize the difference between Inscribed and Circumscribed. Notice that the radius is the same (2") but the Polygons are different sizes. Selecting Inscribed or Circumscribed is important.

INSCRIBED   **CIRCUMSCRIBED**

EDGE METHOD
1. Select the **Polygon** command using one of the options shown above.

2. The following prompts will appear on the command line:

   _polygon Enter number of sides <4>: **type number of sides <enter>**
   Specify center of polygon or [Edge]: **type E <enter>**
   Specify first endpoint of edge: **place first endpoint of edge (P1)**
   Specify second endpoint of edge: **place second endpoint of edge (P2)**

POLYLINES

A **POLYLINE** is very similar to a LINE. It is created in the same way a line is drawn. It requires first and second endpoints. But a POLYLINE has additional features, as follows:

1. A **POLYLINE** is ONE object, even though it may have many segments.
2. You may specify a specific width to each segment.
3. You may specify a different width to the start and end of a polyline segment.

Select the Polyline Command using one of the following:

Ribbon = Home tab / Draw panel / ⟲

Keyboard = PL <enter>

THE FOLLOWING ARE EXAMPLES OF POLYLINES WITH WIDTHS ASSIGNED.

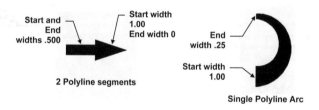

Start and
End
widths .500

Start width
1.00
End width 0

2 Polyline segments

End
width .25

Start width
1.00

Single Polyline Arc

Refer to the next page for more polyline options

POLYLINES....continued

OPTIONS:

WIDTH
Specify the starting and ending width.

You can create a tapered polyline by specifying different starting and ending widths.

HALFWIDTH
The same as Width except the starting and ending halfwidth specifies half the width rather than the entire width.

ARC
This option allows you to create a circular polyline less than 360 degrees. You may use (2) 180 degree arcs to form a full circular shape.

CLOSE
The close option is the same as in the Line command. Close attaches the last segment to the first segment.

LENGTH
This option allows you to draw a polyline at the same angle as the last polyline drawn. This option is very similar to the OFFSET command. You specify the first endpoint and the length. The new polyline will automatically be drawn at the same angle as the previous polyline.

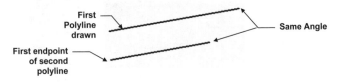

POLYLINES....continued

CONTROLLING THE FILL MODE

<u>If you turn the FILL mode OFF the polylines will appear as shown below.</u>

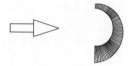

How to turn FILL MODE on or off.
1. Command: *type* **FILL** *<enter>*
2. Enter mode [On / Off] <ON>: *type ON or Off <enter>*
3. Command: *type REGEN <enter> or select: View / Regen*

EXPLODING A POLYLINE

NOTE: If you explode a POLYLINE it loses its width and turns into a regular line as shown below.

EDITING POLYLINES

The **POLYEDIT** command allows you to make changes to a polyline's option, such as the width. You can also change a regular line into a polyline and JOIN the segments.

Note: If you select a line that is **NOT a POLYLINE**, the prompt will ask if you would like to turn it into a POLYLINE.

1. Select the **POLYEDIT** command using one of the following:

    Ribbon = Home tab / Modify ▾ panel /

    Keyboard = PE <enter>

    Note: You may modify "Multiple" polylines simultaneously.

2. PEDIT Select polyline or [Multiple]: *select the polyline to be edited or "M"*

3. Enter an option [Close/Join/Width/Edit vertex/Fit/Spline/Decurve/Ltypegen/Undo]:
    select an Option (descriptions of each are listed below.)

OPTIONS

CLOSE
CLOSE connects the last segment with the first segment of an Open polyline.
AutoCAD considers a polyline open unless you use the "Close" option to connect the segments originally.

Open   Close

OPEN
OPEN removes the closing segment, but only if the CLOSE option was used to close the polyline originally.

EDITING POLYLINES....continued

OPTIONS:

JOIN
The JOIN option allows you to join individual polyline segments into one polyline. The segments must have matching endpoints.

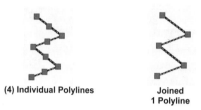

(4) Individual Polylines   Joined
                                 1 Polyline

WIDTH
The WIDTH option allows you to change the width of the polyline. But the entire polyline will have the same width.

EDIT VERTEX
This option allows you to change the starting and ending width of each segment individually.

SPLINE
This option allows you to change straight polylines to curves.

Polyline   Splined

DECURVE
This option removes the SPLINE curves and returns the polyline to its original straight line segments.

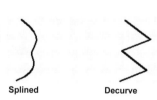

Splined   Decurve

RECTANGLE

A Rectangle is a closed rectangular shape. It is one object not 4 lines.
You can specify the length, width, area, and rotation options.
You can also control the type of corners on the rectangle—fillet, chamfer, or square
and the width of the Line.

First, let's start with a simple Rectangle using the cursor to select the corners.

1. Start the **RECTANGLE** command by using one of the following:

    Ribbon = Home tab / Draw panel /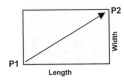

    Keyboard = REC <enter>

2. The following will appear on the command line:

    Command: _rectang
    Specify first corner point or [Chamfer/Elevation/Fillet/Thickness/Width]:

3. Specify the location of the first corner by moving the cursor to a location (**P1**) and
   then press the left mouse button.

    The following will appear on the command line:

    Specify other corner point or [Area / Dimensions / Rotation]:

4. Specify the location of the **diagonal** corner (**P2**) by moving the cursor diagonally
   away from the first corner (**P1**) and pressing the left mouse button.

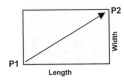

OR

4. Type **D <enter>**

    Specify length for rectangles <0.000>: *Type the desired length <enter>.*

    Specify width for rectangles <0.000>: *Type the desired width <enter>.*

    Specify other corner point or [Dimension]: *move the cursor up, down, right or
    left to specify where you want the second corner relative to the first corner
    and then press <enter> or press left mouse button.*

RECTANGLE....continued

OPTIONS: Chamfer, Fillet and Width

Note: the following options are available **Before** you place the **first corner** of the Rectangle.

CHAMFER

A chamfer is an angled corner. The Chamfer option automatically draws all 4 corners with chamfers simultaneously and all the same size. You must specify the distance for each side of the corner as distance 1 and distance 2.

Example: A Rectangle with dist1 = .50 and dist2 = .25

1. Select the Rectangle command
2. Type C <enter>
3. Enter .50 for the first distance
4. Enter .25 for the second distance
5. Place the first corner (P1)
6. Place the diagonal corner (P2)

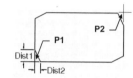

FILLET

A fillet is a rounded corner. The fillet option automatically draws all 4 corners with fillets (all the same size). You must specify the radius for the rounded corners.

Example: A Rectangle with .50 radius corners.
1. Select the Rectangle command
2. Type F <enter>
3. Enter .50 for the radius.
4. Place the first corner (P1)
5. Place the diagonal corner (P2)

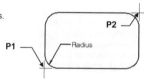

Note: You must set Chamfer and Fillet back to "0" before defining the width. Unless you want fat lines _and_ Chamfered or Filleted corners.

WIDTH

Defines the width of the rectangle lines.
Note: Do not confuse this with the "Dimensions" Length and Width.
Width makes the lines appear fatter.

Example: A Rectangle with a width of .50

1. Select the Rectangle command
2. Type W <enter>
3. Enter .50 for the width.
4. Place the first corner (P1)
5. Place the diagonal corner (P2)

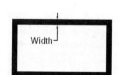

CREATING A REVISION CLOUD

When you make a revision to a drawing it is sometimes helpful to highlight the revision for someone viewing the drawing. A common method to highlight the area is to draw a "Revision Cloud" around the revised area. This can be accomplished easily with the "Revision Cloud" command.

The Revision Cloud command creates a series of sequential arcs to form a cloud shaped object. You set the minimum and maximum arc lengths. (Maximum arc length cannot exceed three times the minimum arc length. Example: Min = 1, Max can be 3 or less) If you set the minimum and maximum different lengths the arcs will vary in size and will display an irregular appearance.

Min & Max same length   **Min & Max different length**

To draw a Revision Cloud you specify the start point with a left click then drag the cursor to form the outline. AutoCAD automatically draws the arcs. When the cursor gets very close to the start point, AutoCAD snaps the last arc to the first arc and closes the shape.

1. Select the Revision Cloud command using one of the following:

   Ribbon = Home tab / Draw ▾ panel /

   Keyboard = Revcloud <enter>

   Command: _revcloud
   Minimum arc length: .50   Maximum arc length: .50   Style: Normal

2. Specify start point or [Arc length/Object/Style] <Object>: ***Select "Arc length"***

3. Specify minimum length of arc <.50>: ***Specify the minimum arc length***

4. Specify maximum length of arc <.50>: ***Specify the maximum arc length***

5. Specify start point or [Arc length/Object/Style] <Object>: ***Place cursor at start location & left click.***

6. Guide crosshairs along cloud path...***Move the cursor to create the cloud outline.***

7. Revision cloud finished. ***When the cursor approaches the start point, the cloud closes automatically.***

CONVERT A CLOSED OBJECT TO A REVISION CLOUD

You can convert a closed object, such as a circle, ellipse, rectangle or closed polyline to a revision cloud. The original object is deleted when it is converted.

(If you want the original object to remain, in addition to the new rev cloud, set the variable "delobj" to "0". The default setting is "1".)

1. Draw a closed object such as a circle.

2. Select the Revision Cloud command using one of the following:

   Ribbon = Home tab / Draw ▾ panel /

   Keyboard = Revcloud <enter>

   Command: _revcloud
   Minimum arc length: .50   Maximum arc length: .50   Style: Normal

3. Specify start point or [Arc length/Object/Style] <Object>: ***Select "Arc length"***

4. Specify minimum length of arc <.50>: ***Specify the minimum arc length***

5. Specify maximum length of arc <.50>: ***Specify the maximum arc length***

6. Specify start point or [Arc length/Object/Style] <Object>: ***Select "Object".***

7. Select object: ***Select the object to convert***

8. Select object: Reverse direction [Yes/No] <No>: ***Select Yes or No***
   Revision cloud finished.

Reverse direction?

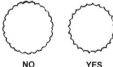

NO   YES

NOTE:
The Match Properties command will not match the arc length from the source cloud to the destination cloud.

REVISION CLOUD STYLE

You may select one of 2 styles for the Revision Cloud; **Normal** or **Calligraphy**.
Normal will draw the cloud with one line width.
Calligraphy will draw the cloud with variable line widths to appear as though you used a chiseled calligraphy pen.

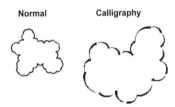

Normal   **Calligraphy**

1. Select the Revision Cloud command using one of the following:

   Ribbon = Home tab / Draw ▾ panel /

   Keyboard = Revcloud <enter>

   Command: _revcloud
   Minimum arc length: .50   Maximum arc length: 1.00   Style: Normal

2. Specify start point or [Arc length/Object/Style] <Object>: ***Select "Style"<enter>***

3. Select arc style [Normal/Calligraphy] <Calligraphy>:***Select "N or C"<enter>***

4. Specify start point or [Arc length/Object/Style] <Object>: ***Select "Arc length"***

5. Specify minimum length of arc <.50>: ***Specify the minimum arc length***

6. Specify maximum length of arc <1.00>: ***Specify the maximum arc length***

7. Specify start point or [Object] <Object>: ***Place cursor at start location & left click.***

8. Guide crosshairs along cloud path...***Move the cursor to create the cloud outline.***

9. Revision cloud finished. ***When the cursor approaches the start point, the cloud closes automatically.***

Section 5
How to…

APPENDIX A
Add a Printer / Plotter

The following are step-by-step instructions on how to configure AutoCAD for your printer or plotter. These instructions assume you are a single system user. If you are networked or need more detailed information, please refer to your AutoCAD Help Index.

Note: You can configure AutoCAD for multiple printers. Configuring a printer makes it possible for AutoCAD to display the printing parameters for that printer.

A. Type: **Plottermanager <enter>**

B. Select **"Add-a-Plotter"** Wizard

Add-A-Plotter Wizard
Shortcut
1 KB

C. Select the **"Next"** button.

D. Select **"My Computer"** then **Next**.

E. Select the **Manufacturer** and the specific **Model** desired then Next.

(If you have a disk with the specific driver information, put the disk in the disk drive and select "Have disk" button then follow instructions.)

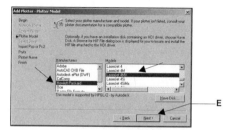

E

F. Select the **"Next"** box.

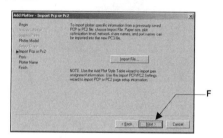

F

G1. Select **"Plot to a port"**.
G2. Then select **"Next"**.

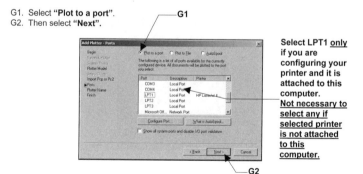

Select LPT1 <u>only</u> if you are configuring your printer and it is attached to this computer. <u>Not necessary to select any if selected printer is not attached to this computer.</u>

G1

G2

5-3

H. The Printer name that you previously selected should appear. Then select **"Next"**

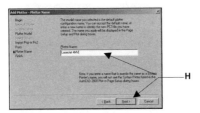

I. Select the **"Edit Plotter Configuration..."** box.

J. Select:
1. Device and Document Settings tab.
2. Media: Source and Size
3. Size: (Select the appropriate size for your printer / plotter)
4. OK box.

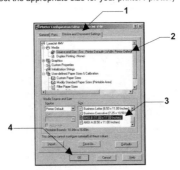

K. Select **"Finish"**.

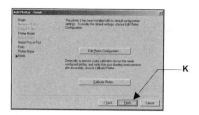

K

L. Type: **Plottermanager <enter>** again.

Is the printer / plotter there?

LaserJet 4MV.pc3
AutoCAD LT Plotter Configura...
2 KB

The new configured printer should
be in the list of printers.

HOW TO CREATE A PAGE SETUP

When you select a layout tab for the first time the Page Setup Manager will appear.
The Page Setup Manager allows you to select the **printer/plotter** and **paper size**.
These specifications are called the "**Page Setup**". This page setup will be saved to
that layout tab so it will be available when ever you use that layout tab.

1. **Open** the drawing you wish to plot.
   (The drawing must be displayed on the screen.)

2. Select a **Layout tab**.

   *Note: If the "Page Setup Manager" dialog box shown below does not appear
   automatically, right click on the Layout tab and select Page Setup Manager.*

3. Select the **New...** button.

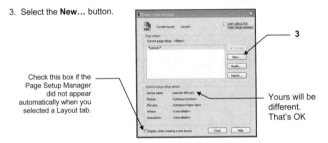

Check this box if the
Page Setup Manager
did not appear
automatically when you
selected a Layout tab.

Yours will be
different.
That's OK

4. Select **<Default output device>** in the Start with: list.

5. Enter the New page setup name: **Setup A**

6. Select **OK** button.

(I am assuming that your
computer is attached to a printer.
If not select Layout1)

Continued on the next page...

HOW TO CREATE A PAGE SETUP....continued

This is where you will select the **printer / plotter**, **paper size** and the **plot offset**.

7. Select the **Printer / Plotter**
   Note: Your current system printer should already be displayed here. If you prefer another select the down arrow and select from the list. If the preferred printer is not in the list you must configure the printer. Refer to Add a Printer in Section 5
8. Select the **Paper Size**

9. Select **Plot Offset**

Notice the name you entered is now displayed as the page setup name.

7 ───
(Yours will be different)

8 ───
(Yours may be 8-1/2 x 11)

9 ───
(Should stay at 0.00000)

10. Select **OK** button.

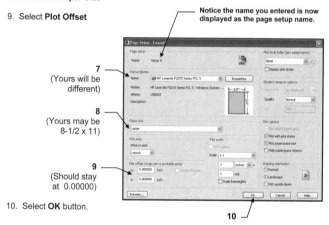

11. Select the Page Setup.

12. Select the **Set Current** button.

13. Select the **Close** button.

11 ───

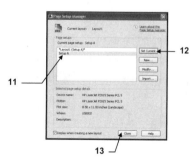

HOW TO CREATE A PAGE SETUP....continued

You should now have a sheet of paper displayed on the screen.
This sheet is the size you specified in the "Page Setup".
This sheet is in front of Model Space.

The dashed line represents the maximum printing area for the printing device that you selected.
Any object outside of this area will not print.

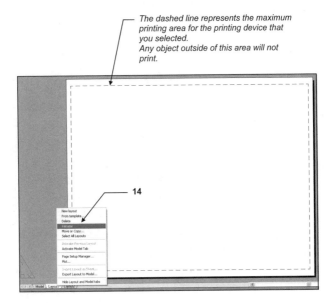

14

Rename the Layout tab

14. Right click on the active Layout tab and select **Rename** from the list.

15. Enter the new Layout name **A Size** <enter>

VIEWPORTS

Note: This is just the concept to get you thinking. The actual step-by-step instructions will follow in the exercises.

Viewports are only used in Paper Space (Layout tab).
Viewports are holes cut into the sheet of paper displayed on the screen in Paper Space.
Viewports frames are objects. They can be moved, stretched, scaled, copied and erased. You can have multiple Viewports, and their size and shape can vary.

Note: It is considered good drawing management to create a layer for the Viewport "frames" to reside on. This will allow you to control them separately; such as setting the viewport layer to "No plot" so it will not be plot.

HOW TO CREATE A VIEWPORT

1. First, create a drawing in <u>Model Space (Model tab)</u> and save it.

2. Select the "Layout1" tab. <u>Layout1 / Layou</u>

   When the "**Page Setup Manager**" dialog box appears, select the Modify button. Then you will select the <u>Printing device</u> and <u>paper size</u> to plot on. (Refer to "How to Create a Page setup")

3. You are now in <u>Paper Space</u>. Model Space appears to have disappeared, because a blank paper is now in front of Model Space, preventing you from seeing your drawing. You designated the size of this sheet in the "page setup" mentioned in #2 above. (The Border, title block and notes will be drawn on this paper.)

You designated the size of this sheet in the Page setup

The dashed line represents the printing limits for the printer that you selected in the Page setup.

Continued on the next page...

VIEWPORTS....continued

4. Draw a border, title block and notes in Paper Space (Layout)

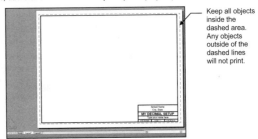

Keep all objects inside the dashed area. Any objects outside of the dashed lines will not print.

Now you will want to see the drawing that is in Model Space.

5. Select layer "**Viewport**"   (You want the viewport frame to be on layer viewport)

6. Type **MV <enter>**

7. Draw a Viewport "frame" by placing the location for the "first corner" and then the "opposite corner" using the cursor. (Similar to drawing a Rectangle, but **do not** use the Rectangle command. You must use the MV command)

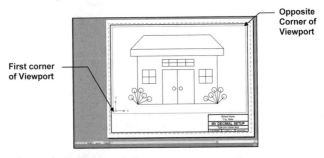

Opposite Corner of Viewport

First corner of Viewport

You should now be able to look through the Paper Space sheet to Model Space and see your drawing because you just cut a rectangular shaped hole in the sheet.

Note: Now you may go back to Model Space or return to Paper Space, simply by selecting the tabs, model or layout.
(Make sure your grids are ON in Model Space and OFF in Paper Space. Otherwise you will have double grids)

HOW TO REACH INTO A VIEWPORT

Here are the rules:
1. You have to be in Paper Space (layout tab) and at least one viewport must have been created.

2. You have to be <u>inside a Viewport</u> to manipulate the scale or position of the drawing that you see in that Viewport.

How to reach into a viewport to manipulate the display.

<u>First, select a layout tab and cut a viewport</u>

At the bottom middle of the screen on the status bar there is a button that either says Model or Paper. This button displays which space you are in currently.

When the button is PAPER you are working on the Paper sheet that is in front of Model Space.
You may cut a viewport, draw border, title block and place notes.

<u>If you want to reach into a viewport to manipulate the display</u>, click inside of the viewport frame. Only one viewport can be activated at one time. The active viewport is indicated by a heavier viewport frame. (Refer to the illustration on the previous page. The viewport displaying the doors is active).
Also, the Paper button changed to Model.

While you are inside a viewport you may manipulate the scale and position of the drawing displayed. To return to the Paper surface click on the word Model and it will change to Paper.   You may now work on the paper surface.

Note: Do not confuse the <u>Model / Layout tabs</u> with the <u>MODEL / PAPER button</u> .

Here is the difference.

The **Model / Layout tabs** shuffle you from the actual drawing area (model space) to the Layout area (paper space). It is sort of like if you had 2 stacked pieces of paper and when you select the Model tab the drawing would come to the front and you could not see the layout. When you select the Layout tab a blank sheet would come to the front and you would not see model space......unless you have a viewport cut.

The **MODEL / PAPER button** allows you to work in model space or paper space without leaving the layout tab. No flipping of sheets. You are either on the paper surface or in the viewport reaching through to model space.

HOW TO LOCK A VIEWPORT

After you have manipulated the drawing within each viewport, to suit your display needs, you will want to **LOCK** the viewport so the display can't be changed accidentally. Then you may zoom in and out and you will not disturb the display.

1. Make sure you are in **Paper Space**.

2. Click on a **Viewport Frame.**

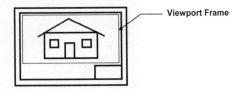

——— **Viewport Frame**

3. Click on the **Open Lock tool** located in the lower right corner of drawing area.

   The icon will change to a **Closed Lock tool**.

   Viewport Unlocked   **Viewport Locked**

Now, any time you want to know if a Viewport is locked or unlocked just glance down to the Lock tool shown above.

Note: Do not use the Lock shown in the lower right corner of the screen.

This locks the size and position of toolbars and windows.

USING THE LAYOUT

Now that you have the correct paper size on the screen, you need to do a little bit more to make it useful.

The next step is to:

1. Add a Border, Title Block and notes in Paper Space.

2. Cut a viewport to see through to Model Space.

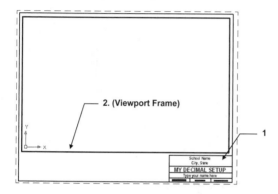

2. (Viewport Frame)

1

School Name
City, State
MY DECIMAL SETUP
Type your name here

3. Adjust the scale of the viewport as follows.
   A. You must be in Paper Space.
   B. Click on the Viewport Frame.
   C. Unlock Viewport if it is locked.
   D. Select the <u>Viewport Scale</u> down arrow ▼.
   E. Select Scale (List shown on next page)

3C

3D

3A

PAPER

Continued on the next page...

USING THE LAYOUT....continued

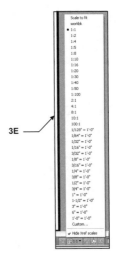

3E

4. Lock the Viewport.

Now you may zoom as much as you desire and it will not affect the adjusted scale.

CREATE A TEMPLATE

1. Set up the drawing and consider the following:
   Drawing Limits
   Units
   Layers
   Text Styles
   Dimension Styles
   Layouts
   Title Blocks

2. Select the **"Application Menu"** ▾

3. Select **"Save As"** ▸

4. Select **"AutoCAD Drawing Template".**

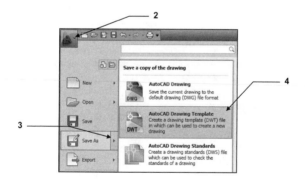

Continued on the next page...

CREATE A TEMPLATE....continued

5. Type the new file name in the **File Name** box. (Example: **2010-Workbook Helper**)
   Do not type the extension .dwt, AutoCAD will add it automatically.

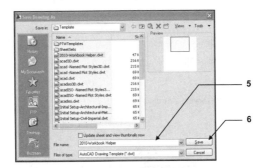

6. Select the **Save** button.

7. Type a description (Example: Use for workbook Lessons 2 through 8)

8. Select **OK** button.

Now you have a template to use for future drawings.

Using a template as a master setup drawing is very good CAD management.

USING A TEMPLATE

TO USE A TEMPLATE

1. Select the **NEW** tool from the **Quick Access Toolbar.**
   1

2. Select **Drawing Template [*.dwt]** from the "<u>Files of type</u>: list.

3. Select the **template name** from the list of templates.

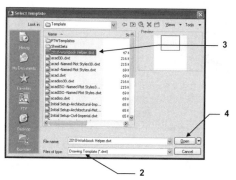

3

4

2

4. Select the **Open** button.

OPENING AN EXISTING DRAWING FILE

Opening an Existing Drawing File means that you would like to open on to the screen a drawing that has been previously created and saved. Usually you are opening it to continue working on it or you need to make some changes.

1. Select the **OPEN** tool on the **Quick Access Toolbar**.

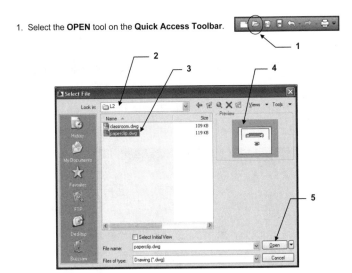

2. Locate the Directory and Folder in which the file had previously been saved.

3. Select the File that you wish to OPEN.

4. A **Thumbnail Preview image** of the file selected is displayed in the Preview area.

5. Select the **Open** button.

MULTIPLE OPEN FILES

AutoCAD allows you to have multiple drawing files open at the same time. If you select the **Application Menu** and select the **Open Documents** icon a list of all open drawings will be displayed.

This is very helpful to **experienced** AutoCAD users but can be confusing for users **NEW** to AutoCAD.

While using this workbook you might find it helpful if you <u>change a setting</u> to prevent opening more than one file at one time. This will only change AutoCAD and will have <u>no affect on any other software on your computer</u>.

1. Close all but one drawing file. You should only have one drawing open.
2. Type: sdi <enter>
3. Type: 1 <enter>

Now AutoCAD will restrict you to one drawing on the screen. If you choose to go back to multiple drawings repeat the steps above except enter "0" instead of "1".

Read-Only files

If the SDI setting is off, allowing multiple open files, and you open a drawing file that you <u>already have open</u> AutoCAD will display this warning.

It is best if you select the **No** button. If you select the **Yes** button, AutoCAD will open a duplicate file as a **"Read-only"** file. A Read-only file should only be used as a reference file. It is not good practice to work on a Read-only file.

SAVING A DRAWING FILE

After you have completed a drawing, it is very important to save it. Learning how to save a drawing correctly is almost more important than making the drawing. If you can't save correctly, you will lose the drawing and hours of work.

There are 2 commands for saving a drawing: **Save** and **Save As**.
I prefer to use **Save As**.

The **Save As** command always pauses to allow you to choose where you want to store the file and what name to assign to the file. This may seem like a small thing, but it has saved me many times from saving a drawing on top of another drawing by mistake.

The **Save** command will automatically save the file either back to where you retrieved it or where you last saved a previous drawing. Neither may be the correct destination. And may replace a file with the same name. So play it safe, use **Save As** for now.

1. Select the **Saveas** command using one of the following:

Quick Access Toolbar =

Application Menu = Save As / AutoCad Drawing

Keyboard = Saveas <enter>

                   2. Select the "Save In" location.
                   (This is where the file will be saved.)

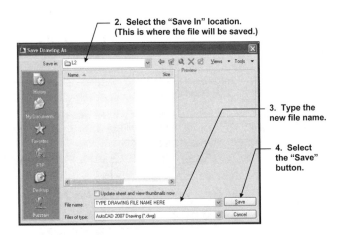

3. Type the new file name.

4. Select the "Save" button.

AUTOMATIC SAVE

AUTOMATIC SAVE

If you turn the automatic save option ON, your drawing is saved at specified time intervals. These temporary files are automatically deleted when a drawing closes normally. The default save time is every 10 minutes. You may change the save time Intervals and where you would prefer the Automatic Save files to be saved.

How to set the Automatic Save intervals

1. Type **options <enter>**

2. Select the **Open and Save** tab.

3. Enter the desired **minutes between saves.**

4. Select the **OK** button.

How to change the Automatic Save location

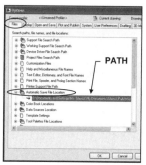

1. Type **options <enter>**

2. Select the **Files** tab

3. Locate the **Automatic Save File Location** and click on the + to display the **path**.

4. Double click on the path

5. Browse to locate the Automatic Save Location desired and highlight it.

6. Select **OK**.

(The browse box will disappear and the new location path should be displayed under the Automatic Save File Location heading)

7. Select **OK** to accept the change.

BACK UP FILES and RECOVER

BACK UP FILES

When you save a drawing file, Autocad creates a file with a **.dwg** extension.
For example, if you save a drawing as **12b**, Autocad saves it as **12b.dwg**. The next time you save that same drawing, Autocad replaces the old with the new and renames the old version **12b.bak**. The old version is now a back up file.
(Only 1 backup file is stored.)

How to open a back up file:
You can't open a **.bak** file.
It must first be renamed with a **.dwg** file extension.

How to view the list of back up files:
The backup files will be saved in the same location as the drawing file.
You must use Windows Explorer to locate the .bak files.

How to rename a back up file:
Right click on the file name.
Select "Rename".
Change the .bak extension to .dwg and press <enter>.

RECOVERING A DRAWING

In the event of a program failure or a power failure open files should be saved automatically.

When you attempt to re-open the drawing the **Drawing Recovery Manager** will display a list of all drawing files that were open at the time of a program or system failure. You can preview and open each .dwg or .bak file to choose which one should be saved as the primary file.

STARTING A NEW DRAWING

Starting A New Drawing means that you want to start with a previously created Template file.

*Note: Do not use the **New** tool if you want to **open** an **existing drawing**.*

HOW TO START A NEW DRAWING

1. Select the **NEW** tool from the **Quick Access Toolbar**.

2. Select the **template file** from the list of templates.

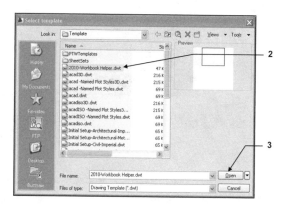

3. Select the **Open** button.

EXITING AUTOCAD

To safely exit AutoCAD follow the instructions below.

1. Save all open drawings.

2. Start the **EXIT** procedure using one of the following.

    Ribbon = None

    Application Menu = Exit AutoCAD

    Keyboard = Exit <enter>

If any changes have been made to the drawing since the last **Save As**, the warning box shown below will appear asking if you want to *SAVE THE CHANGES*?

Select *YES, NO or CANCEL.*

Customizing your Wheel Mouse

A Wheel mouse has two or more buttons and a small wheel between the two topside buttons. The default functions for the two top buttons and the Wheel are as follows:
Left Hand button is for **input** and can't be reprogrammed.
Right Hand button is for **Enter** or the **shortcut menu**.
The Wheel may be used to <u>Zoom and Pan</u> or <u>Zoom and display</u> the Object Snap menu. You will learn more about this later.

The following describes how to select the Wheel functions. After you understand the functions, you may choose to change the setting.
To change the setting you must use the **MBUTTONPAN** variable.

MBUTTONPAN setting 1: (Factory setting)

ZOOM	Rotate the wheel forward to zoom in Rotate the wheel backward to zoom out
ZOOM EXTENTS	Double click the wheel to view entire drawing
PAN	Press the wheel and drag the mouse to move the drawing on the screen.

MBUTTONPAN setting 0:

ZOOM	Rotate the wheel forward to zoom in Rotate the wheel backward to zoom out
OBJECT SNAP	Object Snap menu will appear when you press the wheel

To change the setting:

1. Type: **mbuttonpan <enter>**

2. Enter **0** or **1 <enter>**

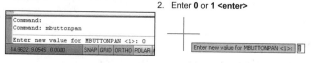

Command:
Command: mbuttonpan
Enter new value for MBUTTONPAN <1>: 0
14.9622, 9.0545, 0.0000   SNAP GRID ORTHO POLAR

Enter new value for MBUTTONPAN <1>: 1

Command Line   **Dynamic Input**

METHODS OF SELECTING OBJECTS

Most AutoCAD commands prompt you to "**select objects**". This means select the objects that you want the command to effect.

There are 2 methods. **Method 1. Pick**, is very easy and should be used if you have only 1 or 2 objects to select. **Method 2. Window**, is a little more difficult but once mastered it is extremely helpful and time saving. Practice the examples shown below.

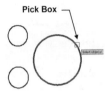

Pick Box ——

Method 1. PICK :
First start a command such as ERASE. (Press E <enter>)
Next you will be prompted to "**Select Objects**", place the
cursor (pick box) on the object. The object will highlight.
This appearance change is called "Rollover Highlighting".
This gives you a preview of which object will be selected.
Press the left mouse button to select the highlighted object.

Method 2. WINDOW: Crossing and Window

Crossing:
Place your cursor in the area <u>up</u> and to the <u>right</u> of the objects
that you wish to select (**P1**) and press the left mouse button.
Then move the cursor down and to the left (**P2**) and press
the left mouse button again.
(Note: The window will be *green* and outer line is *dashed*.)
Only the objects that this window **crosses** will be selected.

In the example on the right, all 3 circles have been selected
because the Crossing Window crosses a portion of each.

Window:
Place your cursor in the area <u>up</u> and to the <u>left</u> of the objects
that you wish to select (**P1**) and press the left mouse button.
Then move the cursor <u>down</u> and to the <u>right</u> of the objects (**P2**)
and press the left mouse button.
(Note: The window will be *blue* and outer line is *solid*.)
Only the objects that this window **completely enclosed** will be
selected.

In the example on the right, only 2 circles have been selected.
(The large circle is ***not*** completely enclosed.)

Section 6
Layers

LAYERS

A **LAYER** is like a transparency. Have you ever used an overhead light projector? Remember those transparencies that are laid on top of the light projector? You could stack multiple sheets but the projected image would have the appearance of one document. Layers are basically the same. Multiple layers can be used within one drawing.

The example, on the right, shows 3 layers.
One for annotations (text), one for dimensions and one for objects.

HOW TO USE LAYERS
First you select the layer and then you draw the objects.
Always select the layer first and then draw the objects.

It is good "drawing management" to draw related objects on the same layer.
For example, in an architectural drawing, you would select layer "walls" and then draw the floor plan.
Then you would select the layer "Electrical" and draw the electrical objects.
Then you would select the layer "Plumbing" and draw the plumbing objects.
Each layer can then be controlled independently.
If a layer is <u>Frozen</u>, it is <u>not visible</u>. When you <u>Thaw</u> the layer it becomes <u>visible</u> again.
(Refer to the following pages for detailed instructions for controlling layers.)

HOW TO SELECT A LAYER

1. Go to **Ribbon = Home tab / Layers panel**

2. Select the drop down arrow ▼

3. Highlight the desired layer and press the left mouse button.

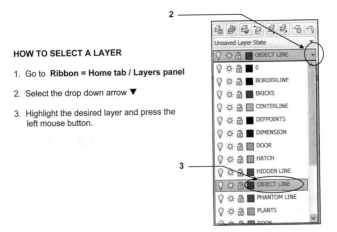

CONTROLLING LAYERS

The following controls can be accessed using the Layer drop down arrow ▼.

ON or OFF
If a layer is **ON** it is **visible**. If a layer is **OFF** it is **not visible**.
Only layers that are **ON** can be **edited** or **plotted**.

FREEZE or THAW
Freeze and **Thaw** are very similar to On and Off. A Frozen layer is <u>not visible</u> and a Thawed layer <u>is visible</u>. Only thawed layers can be edited or plotted.

Additionally:
a. Objects on a Frozen layer **cannot** be accidentally erased
b. When working with large and complex drawings, freezing saves time because frozen layers are not **regenerated** when you zoom in and out.

LOCK or UNLOCK
Locked layers are visible but <u>cannot be edited</u>.
They are visible so they **will** be plotted.

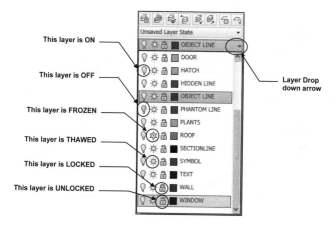

This layer is ON
This layer is OFF
This layer is FROZEN
This layer is THAWED
This layer is LOCKED
This layer is UNLOCKED

Layer Drop down arrow

CONTROLLING LAYERS....continued

To access the following options you must use the **Layer Properties Manager**.
You may also access the options listed on the previous page within this dialog box.

To open the **Layer Properties Manager** use one of the following.

Ribbon = Home tab / Layers panel /

Keyboard = LA <enter>

HOW TO DELETE AN EXISTING LAYER

1. Highlight the layer name to be deleted.
2. Select the **Delete Layer** tool.
Or
1. Highlight the layer name to be deleted.
2. Right click and select **Delete Layer**

PLOT or NOT PLOTTABLE

This tool prevents a layer from plotting even though it is visible within the
Drawing Area.
A Not Plottable layer will not be displayed when using Plot Preview.
If the Plot tool has a slash the layer will not plot.

LAYER COLOR

Color is not merely to make a pretty display on the screen. Layer colors can help define objects. For example, you may assign color Green for all doors. Then, at a glance, you could identify the door and the layer by their color.

Here are some additional things to consider when selecting the colors for your layers.

Consider how the colors will appear on the paper.
(Pastels do not display well on white paper.)

Consider how the colors will appear on the screen.
(Yellow appears well on a black background but not on white.)

How to change the color of a layer.

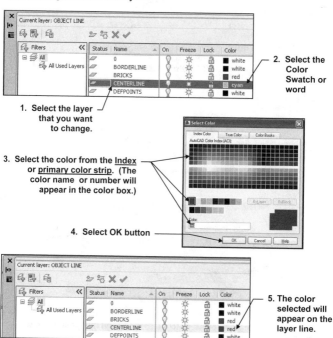

1. Select the layer that you want to change.

2. Select the Color Swatch or word

3. Select the color from the Index or primary color strip. (The color name or number will appear in the color box.)

4. Select OK button

5. The color selected will appear on the layer line.

LINEWEIGHTS

A **Lineweight** means "**how heavy or thin is the object line**".

It is "good drawing management" to establish a contrast in the lineweights of entities.

In the example below the rectangle has a heavier lineweight than the dimensions. The contrast in lineweights makes it easier to distinguish between entities.

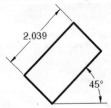

LINEWEIGHT SETTINGS
Lineweights are plotted with the exact width of the lineweight assigned.
But you may adjust how they are displayed on the screen. (Refer to #3 below)

IMPORTANT: Before assigning lineweights you should first select the **Units** and **Display Scale** as shown below.

Select the **Lineweight Settings** box using one of the following:

>   **Ribbon = Not discussed yet**
>
>   **Keyboard = lweight <enter>**
>
>   **Status Bar = Rt. Click on LWT button and**
>   **select Settings.**

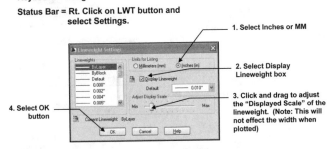

1. Select Inches or MM

2. Select Display Lineweight box

3. Click and drag to adjust the "Displayed Scale" of the lineweight. (Note: This will not effect the width when plotted)

4. Select OK button

NOTE: These settings will be saved to the computer not the drawing and will remain until you change them.

ASSIGNING LINEWEIGHTS

Note: Before assigning **Lineweights** to Layers make sure your **Lineweight settings** (Units and Display scale) are correct. Refer to the previous page.

ASSIGNING LINEWEIGHTS TO LAYERS

1. Select the Layer Properties Manager using one of the following:

   Ribbon = Home tab / Modify panel /

   Keyboard = LA <enter>

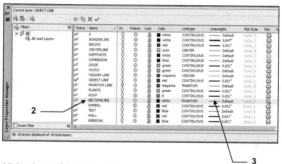

2. Highlight a Layer (Click on the name)

3. Click on the Lineweight for that layer.

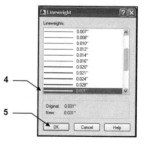

4. Scroll and select a lineweight from the list.

5. Select the **OK** button.

Note: Lineweight selections will be saved within the **current** drawing and will not effect any other drawing.

CREATING NEW LAYERS

Using layers is an important part of managing and controlling your drawing. It is better to have too many layers than too few. You should draw like objects on the same layer. For example, place all doors on the layer "door" or centerlines on the layer "centerline".

When you create a new layer you will assign a **name, color, linetype, lineweight** and whether or not it should **plot**.

1. Select the Layer command using one of the following:

   Ribbon = Home tab / Layers panel /

   Keyboard = LA <enter>

2. New Layer tool

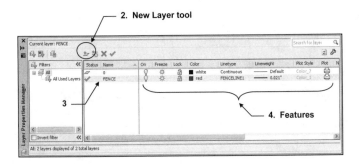

2. Select the **New Layer tool** and a new layer will appear.

3. Type the new layer **name** and press <enter>

4. Select any of the **features** and a dialog box will appear.

LOADING and SELECTING LAYER LINETYPES

In an effort to conserve data within a drawing file, AutoCAD automatically loads only one linetype called "**continuous**". If you would like to use other **linetypes**, such as "dashed" or "fenceline", you must **Load** them into the drawing as follows:

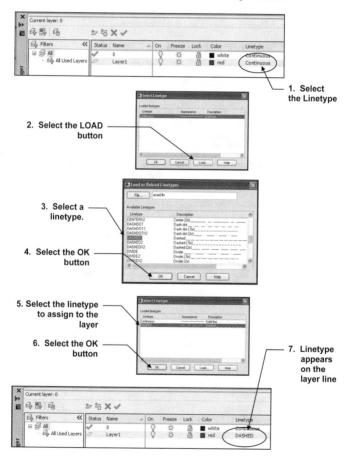

1. **Select the Linetype**

2. **Select the LOAD button**

3. **Select a linetype.**

4. **Select the OK button**

5. **Select the linetype to assign to the layer**

6. **Select the OK button**

7. **Linetype appears on the layer line**

LAYER MATCH

If you draw an object on the wrong layer you can easily change it to the desired layer using Layer Match command. You first select the object that needs to be changed and then select an object that is on the correct layer (object on destination layer).

1. Select the Layer Match command using one of the following:

    Ribbon = Home tab / Layers panel /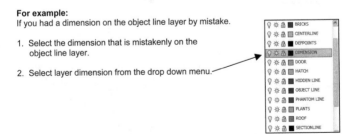

    Browser Menu = Format / Layer Tools / Layer Match

    Keyboard = Laymch <enter>

    Command: _laymch
2. Select objects to be changed:*select the objects that need to be changed*

3. Select objects: *select more objects or <enter> to stop selecting*

4. Select object on destination layer or [Name]: *select the object that is on the layer that you want to change to*

Note:
You may also easily change the layer of an object to another layer as follows:

1. Select the object that you wish to change.

2. Select the Layer drop down menu and select the layer that you wish the object to be placed on.

For example:
If you had a dimension on the object line layer by mistake.

1. Select the dimension that is mistakenly on the object line layer.

2. Select layer dimension from the drop down menu.

Section 7
Input Options

COORDINATE INPUT

In this Section you will learn how to place objects in <u>specific locations</u> by entering coordinates. This process is called **Coordinate Input**.

Autocad uses the ***Cartesian Coordinate System.***
The Cartesian Coordinate System has 3 axes, X, Y and Z.

The **X** is the Horizontal axis. (*Right and Left*)
The **Y** is the Vertical axis. (*Up and Down*)
The **Z** is Perpendicular to the X and Y plane.
(*The Z axis, <u>which is not discussed in this book</u>, is used for 3D.*)

UCS Icon

Look at the User Coordinate System (UCS) icon in the lower left corner of your screen. The X and Y arrows are pointing in the positive direction.

The location where the X , Y and Z axes intersect is called the **ORIGIN**.
*The **Origin** always has a coordinate value of X=0, Y=0 and Z=0 (0,0,0)*
When you move the cursor away from the Origin, in the direction of the arrows, the X and Y coordinates are positive.
When you move the cursor in the opposite direction, the X and Y coordinates are negative.

Using this system, every point on the screen can be specified using positive or negative X and Y coordinates.

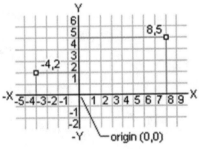

ABSOLUTE COORDINATES

There are <u>3 types of Coordinate input</u>, **Absolute, Relative** and **Polar**.

ABSOLUTE COORDINATES

When inputting absolute coordinates the input format is: **X, Y** (that is: X <u>comma</u> Y)

Absolute coordinates come *from the ORIGIN* and are typed as follows: 8, 5

The first number (8) represents the **X-axis** (horizontal) distance <u>from the Origin</u> and the second number (5) represents the **Y-axis** (vertical) distance from the Origin. The two numbers must be separated by a **comma**.

An absolute coordinate of **4, 2** will be **4** units to the right (horizontal) and **2** units up (vertical) <u>from the current location of the Origin</u>.

An absolute coordinate of **-4, -2** will be **4** units to the left (horizontal) and **2** units down (vertical) <u>from the current location of the Origin</u>.

The following are examples of Absolute Coordinate input. <u>Notice where the Origin is located in each example.</u>

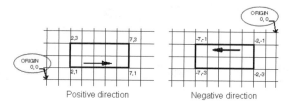

Positive direction   Negative direction

Very important:

While working with coordinates it is best to <u>**turn off**</u> Dynamic Input.

DYN, on the status bar, should be gray.

SNAP | GRID | ORTHO | POLAR | OSNAP | OTRACK | DUCS | DYN | LWT | QP

RELATIVE COORDINATES

RELATIVE COORDINATES

Relative coordinates come *from the last point entered*. (Not from the Origin)

The first number represents the **X-axis** (horizontal) and the second number represents the **Y-axis** (vertical) just like the absolute coordinates.

To distinguish the relative coordinates from absolute coordinates the two numbers must be <u>preceded by an @ symbol</u> in addition to being separated by a **comma**.

A Relative coordinate of **@5, 2** will go to the **right** 5 units and **up** 2 units <u>from the last point entered</u>.

A Relative coordinate of **@-5, -2** will go to the **left** 5 units and **down** 2 units <u>from the last point entered</u>.

The following is an example of Relative Coordinate input.

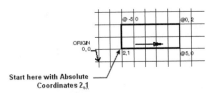

Start here with Absolute
Coordinates 2,1

Very important:

While working with coordinates it is best to **turn off** Dynamic Input.

DYN, on the status bar, should be gray.

SNAP | GRID | ORTHO | POLAR | OSNAP | OTRACK | DUCS | DYN | LWT | QP

EXAMPLES OF COORDINATE INPUT

Scenario 1.
If you want to draw a line with the first endpoint "at the Origin" and the second endpoint 3 units in the positive X direction.

0,0 —————— 3,0

1. Select the Line command.
2. You are prompted for the first endpoint: **Type 0, 0 <enter>**
3. You are then prompted for the second endpoint: **Type 3, 0 <enter>**

What did you do?
The first endpoint coordinate input, 0,0 means that you do not want to move away from the Origin. You want to start "**ON**" the Origin.

The second endpoint coordinate input, 3, 0 means that you want to move 3 units in the positive X axis. The "0" means you do not want to move in the Y axis. So the line will be exactly horizontal.

Scenario 2.
You want to start a line 8 units directly above the origin and it will be 4 units in length, perfectly vertical.

@0,4

1. Select the Line command.
2. You are prompted for the first endpoint: **Type 0, 8 <enter>**
3. You are prompted for the second endpoint: **Type @0, 4 <enter>**

0,8

What did you do?
The first endpoint coordinate input, 0, 8 means you do not want to move in the X axis direction but you do want to move in the Y axis direction.

The second endpoint coordinate input @0, 4 means you do not want to move in the X axis "from the last point entered" but you do want to move in the Y axis "from the last point entered. (Remember the @ symbol is only necessary if you are entering coordinates on the command line.)

Scenario 3.
Now try drawing 5 connecting line segments.
1. Select the Line command.
2. First endpoint: 2, 4 <enter>
3. Second endpoint: @ 2, -3 <enter>
4. Second endpoint: @ 0, -1 <enter>
5. Second endpoint: @ -1, 0 <enter>
6. Second endpoint: @ -2, 2 <enter>
7. Second endpoint: @ 0, 2 <enter> <enter>

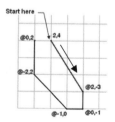

Note: If you enter an incorrect coordinate, just select the undo arrow and the last segment will disappear and you will have another chance at entering the correct coordinate.

DIRECT DISTANCE ENTRY (DDE)

DIRECT DISTANCE ENTRY is a combination of keyboard entry and cursor movement. **DDE** is used to specify distances in the horizontal or vertical axes <u>from the last point entered</u>. **DDE is a *Relative Input*.** Since it is used for Horizontal and Vertical movements, **Ortho must be ON**.

(Note: to specify distances on an angle, refer to Polar Input i

Using DDE is simple. Just move the cursor and type the distance.
Negative and positive is understood automatically by moving the cursor up (positive), down (negative), right (positive) or left (negative) <u>from the last point entered.</u> No minus sign necessary.

Moving the cursor to the right and typing 5 <enter> tells AutoCAD that the 5 is positive and Horizontal.
Moving the cursor to the left and typing 5 <enter> tells AutoCAD that the 5 is negative and Horizontal.
Moving the cursor up and typing 5 <enter> tells AutoCAD that the 5 is positive and Vertical.
Moving the cursor down and typing 5 <enter> tells AutoCAD that the 5 is negative and Vertical.

EXAMPLE:

1. <u>Ortho must be ON</u>.
2. Select the Line command.
3. Type: 1, 2 <enter> to enter the first endpoint using Absolute coordinates.
4. Now move your cursor to the right and type: 5 <enter>
5. Now move your cursor up and type: 4 <enter>
6. Now move your cursor to the left and type: 5 <enter> <enter> to stop

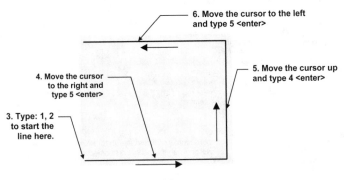

6. Move the cursor to the left and type 5 <enter>

5. Move the cursor up and type 4 <enter>

4. Move the cursor to the right and type 5 <enter>

3. Type: 1, 2 to start the line here.

POLAR COORDINATE INPUT

Previously you learned to control the length and direction of horizontal and vertical lines using Relative Input and Direct Distance Entry. Now you will learn how to control the length and **ANGLE** of a line using **POLAR Coordinate Input**..

UNDERSTANDING THE *"POLAR DEGREE CLOCK"*

Previously when drawing Horizontal and Vertical lines you controlled the direction using a <u>Positive</u> or <u>Negative</u> input. *Polar Input is different*. The Angle of the line will determine the direction.

For example: If I want to draw a line at a 45 degree angle towards the upper right corner, you would use the angle 45. But if I want to draw a line at a 45 degree angle towards the lower left corner, you would use the angle 225.

You may also use Polar Input for Horizontal and Vertical lines using the angles 0, 90, 180 and 270. No negative input is required.

POLAR DEGREE CLOCK

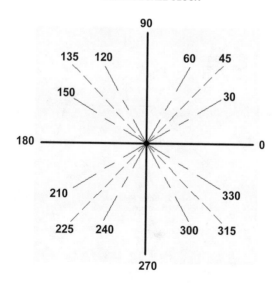

POLAR COORDINATE INPUT....continued

DRAWING WITH *POLAR COORDINATE INPUT*

When entering polar coordinates the first number represents the **Distance** and the
second number represents the **Angle**. The two numbers are separated by the **less
than (<)** symbol. The input format is: **distance < angle**

> Note: If you are using Dynamic Input (DYN button), refer to the next page.
>
> SNAP GRID ORTHO POLAR OSNAP OTRACK DUCS DYN LWT QP

A Polar coordinate of **@6<45** will be a distance of <u>6 units</u> and an angle of <u>45 degrees</u>
from the last point entered.

Here is an example of Polar input for 4 line segments.

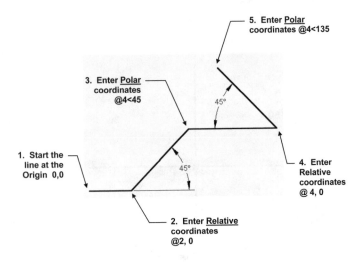

5. Enter <u>Polar</u>
coordinates @4<135

3. Enter <u>Polar</u>
coordinates
@4<45

45°

1. Start the
line at the
Origin 0,0

45°

4. Enter
Relative
coordinates
@ 4, 0

2. Enter <u>Relative</u>
coordinates
@2, 0

POLAR TRACKING

Polar Tracking can be used instead of **Dynamic Input**. When *Polar Tracking* is "**ON**", a dotted *"tracking"* line and a *"tool tip"* box appear. The tracking line.... "snaps" to a **preset angle increment** when the cursor approaches one of the preset angles. The word *"Polar"*, followed by the *"distance"* and *"angle"* from the last point appears in the box.

Tracking Line

Polar: 1.219 < 30°

Polar Tool Tip box

HOW TO SET THE INCREMENT ANGLE

1. Right Click on the **POLAR** button on the Status Bar and select "**SETTINGS**", or select an angle from the list.

| SNAP | GRID | ORTHO | POLAR | OSNAP | OTRACK | DUCS | DYN | LWT | QP |

POLAR ANGLE SETTINGS

Increment Angle   Choose from a list of Angle increments including 90, 45, 30, 22.5, 18, 15,10 and 5. It will also snap to the selected angles multiples. For example: if you choose 30 it will snap to 30, 60, 90, 120, 150, 180, 210, 240, 270, 300, 330 and 0.

Additional Angles   Check this box if you would like to use an angle other than one in the Incremental Angle list. For example: 12.5.

New   You may add an angle by selecting the "New" button. You will be able to snap to this new angle in addition to the incremental Angle selected. But you will not be able to snap to it's multiple. For example, if you selected 7, you would not be able to snap to 14.

Delete   Deletes an Additional Angle. Select the Additional angle to be deleted and then the Delete button.

POLAR ANGLE MEASUREMENT

ABSOLUTE   Polar tracking <u>angles</u> are relative to the UCS.

RELATIVE TO LAST SEGMENT   Polar tracking <u>angles</u> are relative to the last segment.

POLAR SNAP

Polar Snap is used with Polar Tracking to make the cursor snap to specific *distances* and *angles*. If you set Polar Snap distance to 1 and Polar Tracking to angle 30 you can draw lines 1, 2, 3 or 4 units long at an angle of 30, 60, 90 etc. without typing anything. You just move the cursor and watch the tool tips.

SETTING THE POLAR SNAP

1. Set the **Polar Tracking Increment Angle**

2. Right Click on the **SNAP** button on the Status Bar and select "**SETTINGS**"

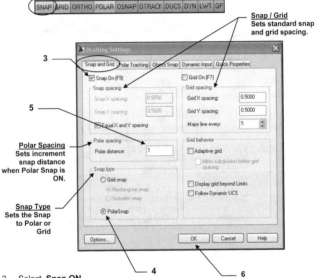

3. Select <u>**Snap ON**</u>

4. Select **PolarSnap**

5. Set the **Polar Distance**

6. Select **OK** button.

DYNAMIC INPUT

To help you keep your focus in the "drawing area", AutoCAD has provided a command interface called **Dynamic Input**. You may input information within the Dynamic Input tool tip box instead of on the command line.

When AutoCAD prompts you for the **First point** the Dynamic Input tool tip displays the (Absolute: X, Y) distance from the Origin.
Enter the **X** dimension, <u>press the Tab key</u>, enter the **Y** dimension then **<enter>**.

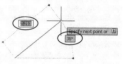

First point

When AutoCAD prompts you for the **Second and Next points** the Dynamic Input tool tip displays the (Relative: Distance and Angle) from the last point entered.
Enter the **distance**, <u>press the tab key</u>, move the cursor in the approximate desired angle and enter the **angle** then **<enter>**.

Second and all next points

How to turn Dynamic Input ON or OFF

Select the **DYN** button on the status bar or use the F12 key.

Here is an example of Dynamic input for 4 line segments.

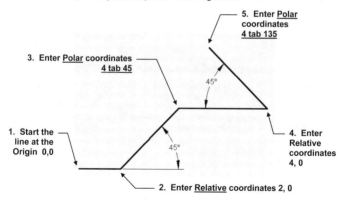

5. **Enter** <u>Polar</u> coordinates <u>4 tab 135</u>

3. **Enter** <u>Polar</u> coordinates <u>4 tab 45</u>

45°

1. **Start the line at the Origin 0,0**

45°

4. **Enter Relative coordinates 4, 0**

2. **Enter** <u>Relative</u> coordinates 2, 0

DYNAMIC INPUT....continued

To enter Cartesian coordinates (X and Y)

1. Enter an "**X**" coordinate value <u>and a **comma**</u>.

2. Enter an "**Y**" coordinate value <enter>.

To enter Polar coordinates (from the last point entered)

1. Enter the **distance** value from the last point entered.

2. Press the **Tab** key.

3. Move the cursor in the approximate direction and enter the **angle** value <enter>

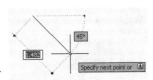

Note: Move the cursor in the approximate direction and enter an angle value of <u>0-180</u> **only** Dynamic Input does not required 360 degrees. (Refer to example on the next page)

How to specify Absolute or Relative coordinates while using Dynamic Input.

<u>To enter **absolute** coordinates</u> when relative coordinate format is displayed in the tooltip. Enter **#** to temporarily override the setting.

<u>To enter **relative** coordinates</u> when absolute coordinate format is displayed in the tooltip. Enter **@** to temporarily override the setting.

Note about Ortho
You may toggle Ortho **On** and **OFF** by holding down the **shift** key.
This is an easy method to use Direct Distance Entry while using Dynamic Input.

Section 8
Miscellaneous

BACKGROUND MASK

Background mask inserts an opaque background so that objects under the text are covered. (masked) The mask will be a rectangular shape and the size will be controlled by the "Border offset factor". The color can be the same as the drawing background or you may select a different color.

1. Select the Multiline Text command.

2. Right click in the Text Area. (The dialog box below appear.)

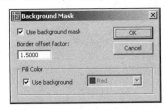

3. Turn this option **ON** by selecting the "Use background Mask" box.

4. Enter a value for the "Border offset factor". The value is a factor of the text height. 1.0 will be exactly the same size as the text. 1.5 (the default) extends the background by 0.5 times the text height. The width will be the same width that you defined for the entire paragraph.

5. In the Fill Color area, select the "Use background" box to make the background the same color as the drawing background. To specify a color, uncheck this box and select a color. (Note: if you use a color you may need to adjust the "draworder" using Tools / Draworder to bring the text to the front.)

BACK UP FILES

When you save a drawing file, Autocad creates a file with a **.dwg** extension.
For example, if you save a drawing as **12b**, Autocad saves it as **12b.dwg**. The next
time you save that same drawing, Autocad replaces the old with the new and renames
the old version **12b.bak**. The old version is now a back up file.
(Only 1 backup file is stored.)

How to view the list of back up files:
Select File / Open
Type "*.bak" in the "file name" box and <enter>. A list of the backup (.bak) files, within
the chosen directory, will appear.

How to open a back up file:
You can't open a **.bak** file.
It must first be renamed with a **.dwg** file extension.

How to rename a back up file:
Right click on the file name.
Select "Rename". Change the .bak extension to .dwg and press <enter>.

GRIPS

Grips are little boxes that appear if you select an object when no command is in use.
Grips must be enabled by typing "grips" <enter> then 1 <enter> on the command line or
selecting the "enable grips" box in the **TOOLS / OPTIONS / SELECTION** dialog box.

Grips can be used to quickly edit objects. You can move, copy, stretch, mirror, rotate,
and scale objects using grips.

The following is a brief overview on how to use three of the most frequently used
options. Grips have many more options and if you like the example below, you should
research them further in the AutoCAD help menu.

1. Select the object (no command can be in use while using grips)
2. Select one of the **blue** grips. It will turn to "**red**". This indicates that it is "**hot**".
   The "**Hot**" grip is the **basepoint**.
3. The editing modes will be displayed on the command line. You may cycle through
   these modes by pressing the SPACEBAR or ENTER key or use the shortcut menu.
4. <u>After editing you must press the ESC key to deactivate the grips on that object.</u>

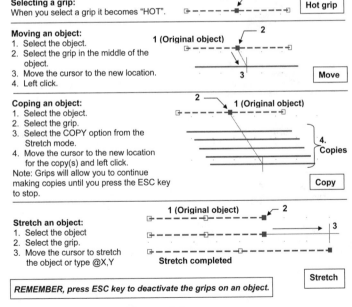

Selecting a grip:
When you select a grip it becomes "HOT".

Hot grip

Moving an object:
1. Select the object.
2. Select the grip in the middle of the
   object.
3. Move the cursor to the new location.
4. Left click.

Move

Coping an object:
1. Select the object.
2. Select the grip.
3. Select the COPY option from the
   Stretch mode.
4. Move the cursor to the new location
   for the copy(s) and left click.
Note: Grips will allow you to continue
making copies until you press the ESC key
to stop.

Copy

Stretch an object:
1. Select the object
2. Select the grip.
3. Move the cursor to stretch
   the object or type @X,Y

Stretch

REMEMBER, press ESC key to deactivate the grips on an object.

8-4

OBJECT SNAP

Increment Snap snaps to increments. So what do you think **Object Snap** snaps to? That's right; "objects". Object snap enables you to snap to "objects" in very specific and accurate locations on the objects.
For example, the endpoint of a line or the center of a circle.

How to select from the Object Snap Menu

1. You must select a command, such as LINE, before you can select Object Snap.

2. While holding down the shift key, press the right mouse button The menu shown below should appear.

3. Highlight and press left mouse button to select.

   Refer to their descriptions on the next page.

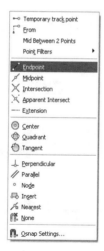

OBJECT SNAP....continued

Object Snap Definitions
Object snap is used when AutoCAD prompts you to place an object. Object snap allows you to place objects very accurately.

A step by step example of "How to use object snap" is shown on the next page.

Note: You may type the **3 bold letters** shown rather than select from the menu.

 ENDpoint — Snaps to the closest endpoint of a Line, Arc or polygon segment. Place the cursor on the object close to the end and the cursor will snap like a magnet to the end of the line.

 MIDpoint — Snaps to the middle of a Line, Arc or Polygon segment. Place the cursor anywhere on the object and the cursor will snap like a magnet to the midpoint of the line.

 INTersection — Snaps to the intersections of any two objects. Place the cursor directly on top of the intersection or select one object and then the other and Autocad will locate the intersection.

 CENter — Snaps to the center of an Arc, Circle or Donut. Place the cursor on the object, or at the approximate center location and the cursor will snap like a magnet to the center.

 QUAdrant — Snaps to a 12:00, 3:00, 6:00 or 9:00 o'clock location on a circle or ellipse. Place the cursor on the circle near the desired quadrant location and the cursor will snap to the closest quadrant.

 PERpendicular — Snaps to a point perpendicular to the object selected. Place the cursor anywhere on the object then pull the cursor away from object and press the left mouse button.

 TANgent — Calculates the tangent point of an Arc or Circle. Place the cursor on the object as near as possible to the expected tangent point.

How to use OBJECT SNAP

The following is an example of attaching a line segment to previously drawn vertical lines. The new line will start from the upper endpoint, to the midpoint, to the lower endpoint.

1. Turn Off **SNAP** and **OSNAP** on the Status Bar.

2. Select the **Line** command.

3. Draw two vertical lines as shown below

4. Select the **Line** command again.

5. Hold the shift key down and press the right mouse button.

6. Select the Object snap **Endpoint** from the object snap menu.

7. Place the cursor close to the upper endpoint of the left hand line.

   The cursor should snap to the end of the line like a magnet. A little red square and a "Endpoint" tooltip are displayed.

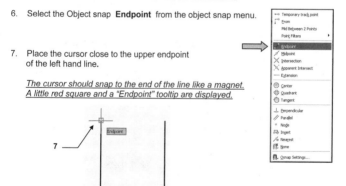

8. Press the left mouse button to attach the new line to the upper endpoint of the previously drawn vertical line. (Do not end the Line command yet.)

Continued on the next page...

How to use OBJECT SNAP....continued

9. Now hold the <u>shift key down</u> and <u>press the right mouse button</u> and select the **Midpoint** object snap option.

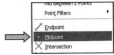

10. Move the cursor to approximately the middle of the right hand vertical line.

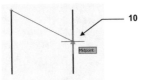

The cursor should snap to the midpoint of the line like a magnet .
A little red triangle with a "Midpoint" tool tip are displayed.

11. Press the left mouse button to <u>attach</u> the new line to the midpoint of the previously drawn vertical line. (Do not end the Line command yet.)

12. Now hold the <u>shift key down</u> and <u>press the right mouse button</u> and select the object snap **Endpoint** again.

13. Move the cursor close to the lower endpoint of the left hand vertical line.

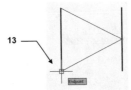

The cursor should snap to the end of the line like a magnet.
A little red square and a tooltip are displayed.

14. Press the left mouse button to <u>attach</u> the new line to the endpoint of the previously drawn vertical line.

15. Stop the Line command and disconnect by pressing **<enter>**.

RUNNING OBJECT SNAP

Selecting Object Snap is not difficult but AutoCAD has provided you with an additional method to increase your efficiency by allowing you to preset select frequently used object snap options. This method is called **RUNNING OBJECT SNAP.**

When **Running Object Snap** is active the cursor will automatically snap to any preset object snap locations thus eliminating the necessity of invoking the object snap menu for each locations.

First you must set the running object snaps and next you must turn ON the Running Object Snap option.

SETTING RUNNING OBJECT SNAP
1. Select the **Running Object Snap** dialog box using one of the following:

    Keyboard = DS <enter>

    Status Bar = Right Click on OSNAP button and select SETTINGS.

2. Select the **Object Snap** tab.

3. Select the Object Snaps desired.
   (In the example below object snap _Endpoint, Midpoint_ and _Intersection_ have been selected.

Note:
Try not to select more than 3 or 4 at one time.
If you select too many the cursor will flit around trying to snap to multiple snap locations. And possibly snap to the wrong location. You will lose control and it will confuse you.

4. Select the **OK** button.

5. Turn <u>ON</u> the **OSNAP** button on the <u>Status Bar</u>.

| SNAP | GRID | ORTHO | POLAR | OSNAP | OTRACK | DUCS | DYN | LWT | QP |

PAN

After you zoom in and out or adjust the scale of a viewport the drawing within the viewport frame may not be placed as you would like it. This is where **PAN** comes in handy. **PAN** will allow you to move the drawing around, within the viewport, without affecting the size.

Note: Do not use the MOVE command. You do not want to actually move the original drawing. You only want to slide the viewport image, of the original drawing, around within the viewport.

How to use the PAN command.

1. Select a layout tab (paper space)

2. Unlock the viewport if it is locked.

3. Click inside a viewport.

4. Select the **PAN** command using one of the following:

   Ribbon = View tab / Navigate panel /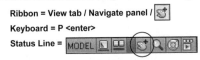
   Keyboard = P <enter>
   Status Line =

4. Place the cursor inside the viewport and hold the left mouse button down while moving the cursor. (Click and drag) When the drawing is in the desired location release the mouse button.

5. Lock the viewport.

Refer to the Example on the next page.

PAN....continued

Before PAN

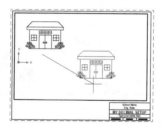

Click once in viewport
to activate it.

To use PAN
(Click, Drag, Release)

After PAN

PROPERTIES PALETTE

The **Properties Palette**, shown below, makes it possible to change an object's properties. You simply open the Properties Palette, select an object and you can change any of the properties that are listed.

How to open the Properties Palette:
    Ribbon = Home tab / Properties panel / ↘

    Keyboard = Ctrl + 1

(An example of how to use the Properties Palette is on the next page.)

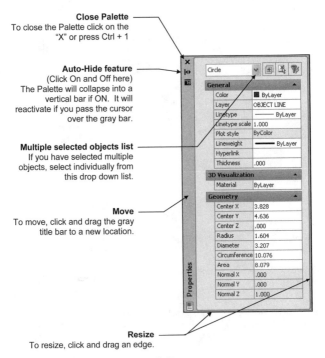

Close Palette
To close the Palette click on the "X" or press Ctrl + 1

Auto-Hide feature
(Click On and Off here)
The Palette will collapse into a vertical bar if ON. It will reactivate if you pass the cursor over the gray bar.

Multiple selected objects list
If you have selected multiple objects, select individually from this drop down list.

Move
To move, click and drag the gray title bar to a new location.

Resize
To resize, click and drag an edge.

PROPERTIES PALETTE....continued

Example of editing an object using the Properties Palette

1. Draw a 2.00 Radius circle.

2. Open the Properties Palette and select the Circle. (*The Properties for the Circle should appear. You may change any of the properties listed in the Properties Palette for this object. When you press <enter> the circle will change.*)

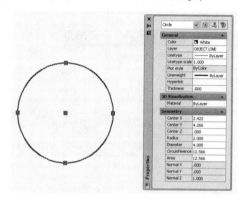

3. Highlight and <u>change the "Radius" to 1.00</u> and the "<u>Layer" to HIDDEN LINE</u> <enter>. *The Circle got smaller and the Layer changed as shown below.*

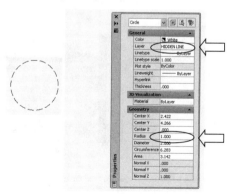

QUICK PROPERTIES PANEL

The Quick Properties Panel, shown below, will only appear if you have it set to ON.
The Quick Properties Panel displays fewer properties and appears when you click
once on an object. You may make changes to the objects properties using Quick
Properties just as you would using the Properties Palette. (AutoCAD is just giving you
another option)

How to turn Quick Properties Panel On or OFF.
Select the **QP** button on the status bar.

Select an object

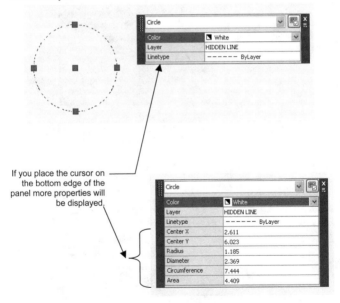

If you place the cursor on
the bottom edge of the
panel more properties will
be displayed.

Refer to the next page to "Customize" the Quick Properties Panel.

8-14

CUSTOMIZING THE QUICK PROPERTIES PANEL

You may add or remove properties from the Quick Properties Panel. <u>And it is easy</u>.

1. Select the **Customize** button.

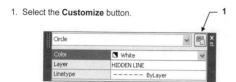

2. Select the Object, from the list, that you would like to customize.

3. <u>Check</u> the boxes for the properties that you <u>want to appear</u>.
   <u>Uncheck</u> the boxes that you <u>do not want to appear</u>.

4. Select **OK**

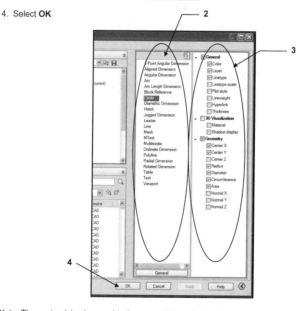

Note: The customizing is saved to the computer not the drawing file.

Notes:

Section 9
Plotting

BACKGROUND PLOTTING

Background Plotting allows you to continue to work while your drawing is plotting. This is a valuable time saver because some drawings take a long time to plot. Or maybe you have multiple drawings to plot and you do not want to tie up your computer.

If you wish to view information about the status of the plot, click on the plotter icon located in the lower right corner of the AutoCAD window.

When the plot is complete, a notification bubble will appear.

 If you do not want this bubble to appear you may turn it off. Right click on the plotter icon and select "**Enable Balloon Notification**". This will remove the check mark and turn the notification off. You may turn it back on using the same process.

When you click on the "Click to view plot and publish details..." , on the bubble, you will get a report like the one shown below. The report will list details about all of the drawings plotted in the current drawing session.

 You may turn Background Plotting ON or OFF. The default setting is OFF.
To turn it ON or OFF:
select **Tools / Options / Plot and Publish tab.**

PLOTTING FROM MODEL TAB

1. **Important:** Open the drawing you want to plot.
2. Make sure that the Model tab is selected.   (Model tab)
3. Select: View / Zoom / All
4. Select the **Plot** command by "right clicking" on the "Model" tab or using one of the following methods listed below:

> **Type = Print or Plot**
> **Pulldown = File / Plot**
> **Tool bar = Standard**

The Plot dialog box below should appear.

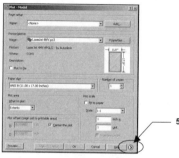

5. " More Options" button

5. Select the "**More Options**" button to expand the dialog box.

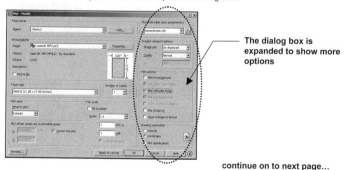

The dialog box is expanded to show more options

continue on to next page...

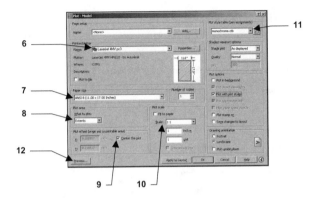

6. Select a "Printer / Plotter" such as the **LaserJet 4MV** from the drop down list.
   Note: This Printer represents a size 17 X 11. If you do not find your printer in the drop down list, you need to configure it on your system.
   Refer to "Configuring a Plotter".

7. Select the Paper Size such as **ANSI B (11 X 17 inches)**

8. Select the Plot Area **EXTENTS**

9. Select the Plot Offset **Center the plot**

10. Select Plot Scale **1 : 1**
    (Note: If you would like to print your drawing on a 8-1/2 X 11 printer, select the printer and change the scale to "Fit to Paper" and change paper size.)

11. Select the Plot Style table **None** for <u>color</u> or **Monochrome.ctb** for <u>Black only</u>.
    The following box will appear. Select **Yes**

Continue on to next page...

12. Select **Preview** button.

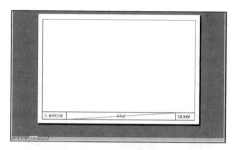

Does your display appear as correct?
If yes, press <enter> and proceed to 13.
If no, recheck 1 through 11.

You have just created a **Page Setup**. All of the settings you have selected can now be saved. You will be able to recall these settings for future plots using this page setup. To save the **Page Setup** you need to **ADD** it to the model tab within this drawing.

13. Select the ADD button.   — 13

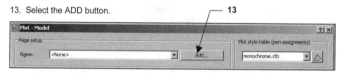

14. Type the new Page Setup name such as: **Model-mono**
(This name identifies that you will use it when plotting the Model tab in monochorme.)

15. Select **OK** button

continue on to next page....

16. Select **Apply to Layout** button.

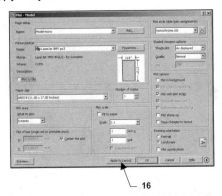

16

17. Select the **OK** button to send the drawing to the printer or select **Cancel** if you do not want to print the drawing at this time. The Page Setup will still be saved.

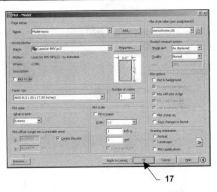

17

18. Save the entire drawing again. The Page Setup will be saved to the **Model tab** within the drawing and available to select in the future. You will not have to select all the individual settings.

Note: The instructions above are to be used when plotting from the "Model" tab. To plot from a Layout tab refer to Plotting from the Layout tab.

HOW TO PLOT FROM PAPER SPACE

STEP 1. SPECIFY PLOT SETTINGS

This means:
- Verify the Plot device and Paper size.
- Select what area of the drawing to plot, what scale to use, where to place the drawing on the paper and which plot style table to use.

A. Open the drawing you wish to plot.

B. Select the layout tab you wish to plot.

C. Select **File / Plot** or place the cursor on the layout tab and press the right mouse button and select **Plot** from the short cut menu.

The Plot dialog box shown below should appear.
Select the "More Options" button in the lower right corner
if your dialog box does not appear the same as shown below

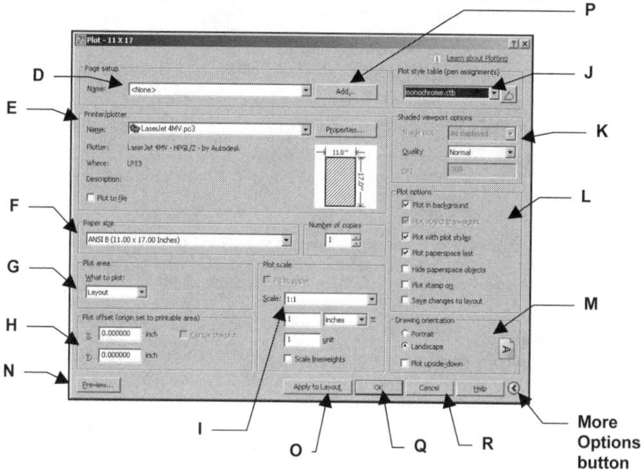

D. **Page Setup name**:
Notice that currently there is no name specified for the Page Setup. You will add one after you have made all of the selections.
If you have previously created a page setup you may select from the drop down list.

E. **Printer / Plotter:**
Verify the **Name.** All previously configured devices will be listed here.

F. **Paper Size:**
Verify the **Paper size**.
The paper sizes shown in the drop down list are the available sizes for the printer that you selected. If the size you require is not listed the printer you selected may not be able to handle that size. For example, a letter size printer cannot handle a 24 X 18 size sheet. You must select a large format printer.

G. **Plot Area:**
Select the area to plot. Layout is the default.

Note: If you selected printer <u>HP4MV</u> select "Layout".
If you selected <u>your printer</u>, use "Extents".

Limits plots the area inside the drawing limits.
(This option is only available when plotting from model space)

Layout plots the paper size
(Select when plotting a Layout)

Extents plots all objects in the drawing file even if out of view.
(This option only available if you have a viewport)

Display plots the drawing as displayed on the screen.

Window plots objects inside a window. To specify the window, choose **Window** and specify the first and opposite (diagonal) corner of the area you choose to plot. (Similar to the Zoom / Window command)

H. **Plot offset:**
The plot can be moved away from the lower left corner by changing the X and/or Y offset.
<u>If you select Plot area "Display" or "Extents", select "**Center the plot.**"</u>

I. **Scale:** Select a **scale** from the drop down list or enter a custom scale.
Note: If you selected printer HP4MV, select scale 1 : 1.
If you selected your printer, select "<u>Fit to Paper size</u>".

*Note: This scale is the Paper Space scale. The Model space scale will be adjusted within the viewport. If you are plotting from a "**LAYOUT**" tab, normally you will use*
__plot scale 1:1__.

J. **Plot Style Table:** Select the Plot Style Table from the list. The Plot Styles determine if the plot is in color, Black ink or screened. You may also create your own.

K. **Shaded viewport options**
This area is used for printing shaded objects when using 3D and will be discussed in the Advanced workbook.

L. **Plot options**
Plot Object Lineweights = plots objects with assigned lineweights.

Plot with Plot Styles = plots using the selected Plot Style Table.

Plot paperspace last = plots model space objects before plotting paperspace objects. Not available when plotting from model space.

Hide Paperspace Objects = used for 3D only. Plots with hidden lines removed.

Plot Stamp = Allows you to print information around the perimeter of the border such as; drawing name, layout name, date/time, login name, device name, paper size and plot scale.

Save Changes to Layout = Select this box if you want to save all of these settings to the current Layout tab.

M. **Drawing Orientation**.
Portrait = the short edge of the paper represents the top of the page.

Landscape = the long edge of the paper represents the top of the page

Plot Upside-down = Plots the drawing upside down.

N. Select **Preview** button.
Preview displays the drawing as it will plot on the sheet of paper.

(Note: If you cannot see through to Model space, you have not cut your viewport yet)

If the drawing is centered on the sheet, press the **Esc** key and continue.

If the drawing does not look correct, press the **Esc** key and check all your settings, then preview again.

Note: If you set your Viewport layer to "No Plot" the Viewport "frame" will not appear when you select Preview. What you see is what will print on the paper.

O. **Apply to Layout**
   This applies all of the previous settings to the layout tab. Whenever you select this layout tab the settings will be already set for you.

P. **Save the Page Setup**
   At this point you have the option of saving these settings as another page setup for future use, not just this layout tab. If you wish to save this setup, select the **ADD** button, type a name and select **OK.**

Q. If your computer **is** connected to the plotter / printer selected, select the **OK** button to plot, then proceed to **S**.

R. If your computer is **not** connected to the plotter / printer selected, select the **Cancel** button to close the Plot dialog box and proceed to **S**.

Note: Selecting Cancel <u>will cancel</u> your selected setting if you did not save the page setup as specified in **P** above.

S. Save the drawing
   This will guarantee that the Page Setup you just created will be saved to this file for future use.

Section 10
Settings

DRAWING SETUP

When drawing with a computer, you must "set up your drawing area" just as you would on your drawing board if you were drawing with pencil and paper. You must decide what size your paper you will need, what Units of measurement you will use (feet and inches or decimals, etc) and how precise you need to be. In CAD these decisions are called "Setting the **Drawing Limits, Units** and **Precision**".

AutoCAD 2010 also has a new feature "Initial Setup" that allows you to select an Industry, add task-based tools and specify the drawing template you want to use when creating a new drawing. See Help Menu for detailed instructions.

DRAWING LIMITS

Consider the drawing limits as the size of the paper you will be drawing on.
You will first be asked to define where the lower left corner should be placed, then the upper right corner, similar to drawing a Rectangle. An 11 x 8.5 piece of paper would have a **lower left corner** of 0,0 and an **upper right corner** of 11, 8.5. *(11 is the horizontal measurement X-axis and 8.5 is the vertical measurement Y-axis.)*

HOW TO SET THE DRAWING LIMITS

1. Select the **DRAWING LIMITS** command by typing: **Limits <enter>**

2. The following will appear on the command line:

   Command: '_limits
   Reset Model space limits:
   Specify lower left corner or [ON/OFF] <0.000,0.000>:

   Note: Limits default setting is ON. AutoCAD will remind you when you are drawing outside of the Limits. If you change the limits to Off you may draw outside of the Limits.

3. Type the X,Y coordinates **0, 0 <enter>** for the lower left corner location of your piece of paper .

4. The command line should now read:

   Specify upper right corner <11.000, 8.5>:

   ```
               11,8.5
     0,0
   ```

5. Type the X,Y coordinates **11, 8.5 <enter>** for the upper right corner of your piece of paper .

6. **This next step is very important**:

   Type **Z <enter> A<enter>** to make the screen display the new drawing limits.
   (This is the shortcut for Zoom / All)

DRAWING SETUP continued....

Grids within Limits

If you have your **Grid behavior** setting **Display grid beyond Limits** turned Off
(no check mark) the grids will only be displayed within the Limits that you set.

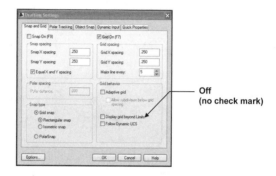

Off
(no check mark)

Grids displayed within
Limits only.

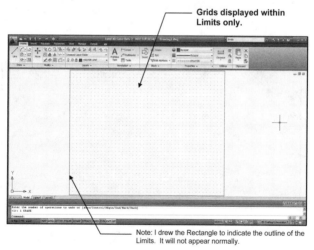

Note: I drew the Rectangle to indicate the outline of the
Limits. It will not appear normally.

DRAWING SETUP continued....

Grids beyond Limits

If you have your **Grid behavior** setting **Display grid beyond Limits** turned **ON** (check mark) the grids will be displayed beyond the Limits that you set.

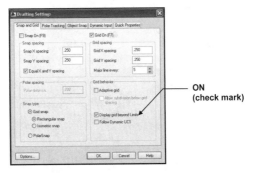

ON
(check mark)

Grids displayed beyond Limits.

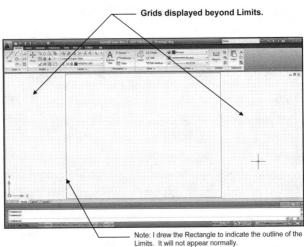

Note: I drew the Rectangle to indicate the outline of the Limits. It will not appear normally.

DRAWING SETUP....continued

UNITS AND PRECISION
You now need to select what unit of measurement with which you want to work.
Such as: Decimal (0.000) or Architectural (0'-0").
Next you should select how precise you want the measurements. For example, do you
want the measurement rounded off to a 3 place decimal or the nearest 1/8".

HOW TO SET THE UNITS AND PRECISION.

1. Select the **UNITS** command using one of the following:

    Application Menu = Drawing / Units [0.0]

    Keyboard = Units <enter>

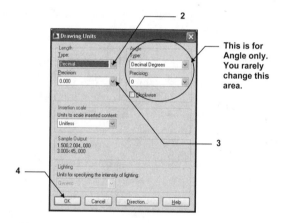

2. **Type:** Select the down arrow and select : **Decimals** or **Architectural**.

3. **Precision:** Select the down arrow and select the appropriate **Precision**
   associated with the "type".
   Examples: 0.000 for Decimals or 1/16" for Architectural.

4. Select the **OK** button to save your selections.

   Easy, yes?

ANNOTATIVE PROPERTY

In this Section you will learn to create a new dimension style and text style and include the **Annotative property** in both.

The Annotative property automates the process of scaling text, dimensions, hatch, tolerances, leaders and symbols. The height of these objects will automatically be determined by the annotation scale setting.

For now I just want you to know how to select it when creating your new styles.

TEXT STYLE

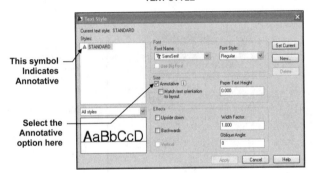

This symbol Indicates Annotative

Select the Annotative option here

DIMENSION STYLE

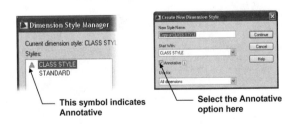

This symbol indicates Annotative

Select the Annotative option here

ANNOTATIVE OBJECTS

You can add the **"Annotative" property** to Text or Dimensions simply by placing a check mark in the Annotative box.

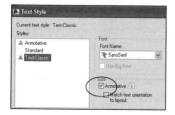 

The Text style and Dimension Style shown above are now **Annotative**.
Notice the Annotative symbol ⚜ beside the Name. (No symbol by Standard name.)

"Annotative objects" are scaled automatically to match the scale of the viewport.
For example, if you want the text, inside the viewport, to print .200 in height and the viewport scale is 1:2 AutoCAD will automatically scale the text height to .400. The text height needs to be scaled by a factor of 2 to compensate for the model space contents appearing smaller.

The easiest way to understand how **Annotative property** works is to do it.
So try this example.

1. Start a NEW file.

2. Select the **Model** tab.
3. Draw a rectangle 3.00 Lg X 1.50 wide   Use layer Object line.

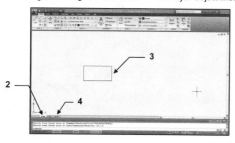

4. Select a **Layout** tab.

Continued on the next page...

ANNOTATIVE OBJECTS....continued

5. Erase any existing viewport frames.

6. Cut 2 new viewports as shown below. Use Layer Viewport

7. Activate the left viewport (double click inside the viewport frame) and do the following:
   A. Adjust the scale of the viewport to **1 : 1**
   B: PAN to place rectangle in the center of the viewport.
   C. Lock the viewport

8. Activate the right viewport (click inside the viewport frame) and do the following:
   A. Adjust the scale of the viewport to **1 : 4**
   B: PAN to place rectangle in the center of the viewport.
   C. Lock the viewport

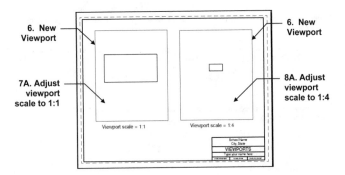

Your screen should appear approximately like this.

Continued on the next page...

ANNOTATIVE OBJECTS....continued

9. Activate the left hand viewport. (Double click inside left hand viewport)

10. Add .200 height text **Scale 1:1** as shown using text style "Text-Classic".
    (Note: Text style "Text-Classic" is annotative)

11. Add the dimension shown using dimension style "Dim-Decimal".
    (Note: Dimension style "Dim-Decimal" is annotative)

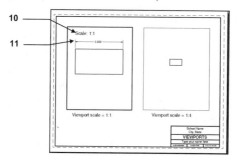

12. Activate the right hand viewport. (Click inside right hand viewport)

13. Add .200 height text **Scale 1:4** as shown using text style "Text-Classic".
    (Note: Text style "Text-Classic" is annotative)

14. Add the dimension shown using dimension style "Dim-Decimal".
    (Note: Dimension style "Dim-Decimal" is annotative)

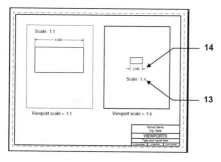

Continued on the next page...

ANNOTATIVE OBJECTS....continued

Notice that the text and dimensions appear the <u>same size in both viewports</u>.

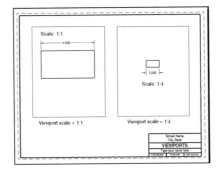

15. Now select the **Model** tab.

Notice there is 2 sets of text and dimensions.

One set has the annotative scale of 1:1 and will be visible only in a 1:1 viewport.
One set has the annotative scale of 1:4 and will be visible only in a 1:4 viewport.
But you see both sets when you select the model tab.

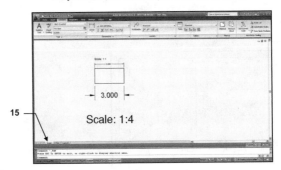

Continued on the next page...

ANNOTATIVE OBJECTS....continued

16. Place your cursor on any of the text or dimensions. An "Annotative symbol"
    will appear. This indicates that this object is annotative and it has only one
    annotative scale.

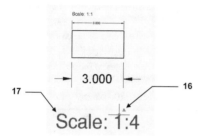

17. Click once on the **Scale: 1:4** text. The Quick properties box should appear.

Notice the text height is listed twice.

Paper text height = .200   This is the height that you selected when placing the text.
When the drawing is printed the <u>text will print .200</u>.

Model text height = .800   This is the desired height of the text (.200) factored by the
viewport scale (1:4). The viewport scale is a factor of 4. (4 X .200 = .800)

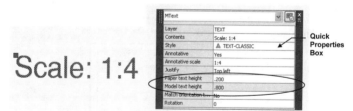

Summary:
If an object is annotative AutoCAD automatically adjusts the scale of the object to the
viewport scale. The most commonly used Annotative Objects are: Dimensions, Text,
Hatch and Multileaders. Refer to the Help menu for more.

ASSIGNING MULTIPLE ANNOTATIVE SCALES

The previous pages showed you how annotative text and dimensions are automatically scaled to the viewport scale. But in order to have an annotative object appear in both viewports you placed 2 sets of text and dimensions. Now you will learn how to easily assign multiple annotative scales to a single text string or dimension so you need not duplicate them each time you create a new viewport. You will just assign an additional annotative scale to the annotative object.

Again, the easiest way to understand this process it to do it. The following is a step by step example.

1. Use the example from the previous pages. *If you did not complete the example from the previous pages, go back and do it now.*

2. Make the right hand viewport active. (Double click inside the viewport)

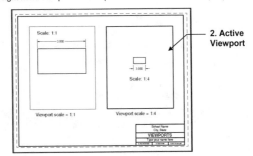

3. Erase the text and the dimension in the right hand viewport only.

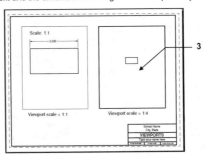

Continued on the next page...

ASSIGNING MULTIPLE ANNOTATIVE SCALES continued

4. Display all annotative objects in all viewports as follows:
   A. Select the **Annotation Visibility** button located in the lower right corner of the drawing status bar.

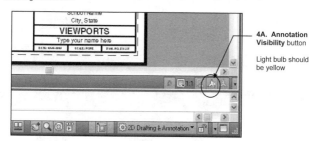

4A. Annotation Visibility button

Light bulb should be yellow

ON (<u>Yellow</u> light bulb) Displays **all** Annotation objects in **all** viewports.
(example below)

OFF (<u>Blue</u> light bulb) Displays **only** Annotation objects that have an annotative scale that matches the viewport scale.

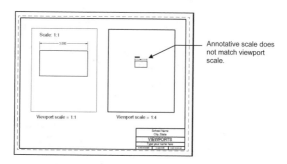

Annotative scale does not match viewport scale.

The dimensions and text are now displayed in both viewports. But the annotative scale of the dimensions and text in the right hand viewport do not match the scale of the viewport. (Notice they are smaller) **The scale of annotative objects must match the scale of the viewport.** _Follow the steps on the next page to assign multiple annotative scales to an annotative object._

Continued on the next page...

ASSIGNING MULTIPLE ANNOTATIVE SCALES continued

5. Place your cursor near the **dimension** in the right hand viewport. Notice the <u>single</u> **Annotation symbol.** This single symbol indicates the annotative dimension has only one annotative scale assigned to it.

6. Select the **Annotate** tab.

7. Select <u>only</u> the <u>dimension</u> in the <u>right hand viewport</u>.

8. Select the **Add Current Scale** tool on the <u>Annotation Scaling</u> panel.

The annotative dimension should have increased in size as shown below.

Increased in size

Scale : 1:1

3.000

9. <u>Turn OFF</u> the **Annotation Visibility.** (Click on button. The light should turn blue)

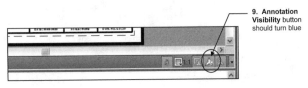

9. **Annotation Visibility** button should turn blue

Continued on the next page...

*Notice the text in the right hand viewport is no longer visible. When the **Annotative Visibility** is OFF only the annotative objects that match the viewport scale will remain visible. The dimension is the only annotative object that has the 1:4 annotative scale assigned to it.*

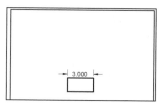

Annotative Visibility OFF

10. Place your cursor near the dimension. Notice 2 annotation symbols appear now. This indicates 2 annotation scales have been assigned to the annotative object.

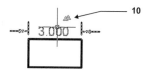

11. Click on the dimension to display the grips and drag the dimension away from rectangle approximately as shown below. (The dimension was too close to the rectangle) <u>Notice the dimension in the left hand viewport did not move.</u> They can be moved individually.

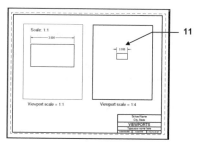

HOW TO REMOVE AN ANNOTATIVE SCALE

If you have an annotative object such as a dimension, that you would like to remove from a viewport, you must remove the annotative scale that matches the viewport scale. **Do not delete** the dimension because it will also be deleted from all of the other viewports. This sounds complicated but is very easy to accomplish.

Problem: I would like to remove **dimension A** from the right hand viewport but I do not want to remove **dimension B** in the lower viewport.

Solution: I must remove the 1/4" = 1" annotation scale from **dimension A**.

(Step by step instructions below.)

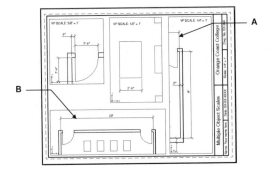

Step 1.

1. Select the **Annotate** tab.

2. Select **dimension A** shown above. (You must be inside the viewport)

3. Select **Add / Delete Scales** tool located on the **Annotation Scaling** panel.

Continued on the next page...

HOW TO REMOVE AN ANNOTATIVE SCALE continued

4. Select the annotation scale to remove. (1/4" = 1')

5. Select the **Delete** button.
   (Remember, you are deleting the annotative scale not the actual dimension. The dimension still exists but it will not have an annotative scale of 1/4"=1' assigned to it. As a result it will not be visible within any viewport that has been scaled to 1/4" = 1')

6. Select the **OK** button.

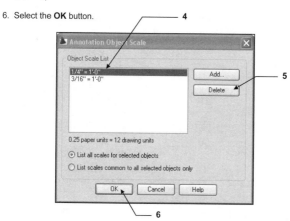

Note: **Dimension A** has been removed and **dimension B** remains.

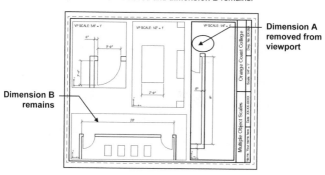

ANNOTATIVE HATCH

When you add cross-hatching to a drawing it is usually in just one area or viewport. But if you would like it shown in all viewports follow the steps below.

1. Draw the hatch in one of the viewports using **<u>Annotative</u>** hatch.
   a. Select the **Annotative** box

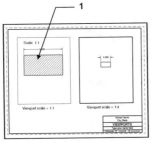

2. Turn ON **Annotation Visibility**

3. Select the <u>Hatch Set </u>to change.

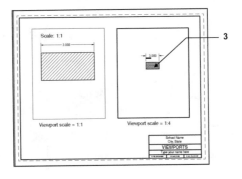

ANNOTATIVE HATCH

4. Select the "**Add Current Scale**" tool located on the **Annotate** tab, **Annotation Scaling** panel.

5. Turn OFF **Annotation Visibility**

Now the appearance of the hatch sets should be identical in both viewports.

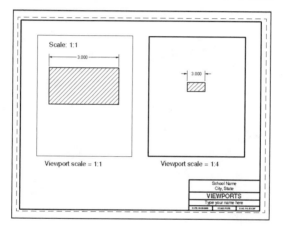

Notes:

Section 11
Text

CREATING NEW TEXT STYLES

AutoCAD provides you with two preset Text Styles named "Standard" and "Annotative".
You may want to create a new text style with a different font and effects.
The following illustrates how to create a new text style.

1. Select the TEXT STYLE command using one of the following:

   Ribbon = Annotate tab / Text panel / ↘

   Keyboard = ST <enter>

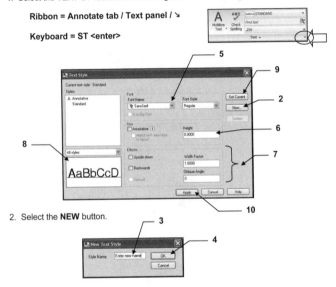

2. Select the **NEW** button.

3. Type the new style name in **STYLE NAME** box.
   Styles names can have a maximum of 31 characters, including letters, numbers,
   dashes, underlines and dollar signs. You can use Upper or Lower case.

4. Select the **OK** button.

5. Select the **FONT**.

6. Enter the value of the Height.
   **Note: If the value is 0, AutoCAD will always prompt you for a height.
   If you enter a number the new text style will have a fixed height and AutoCAD
   will not prompt you for the height.**

CREATING NEW TEXT STYLES....continued

7. Assign **EFFECTS**.

UPSIDE-DOWN
Each letter will be created upside-down in the order in which it was typed.
(Note: this is different from rotating text 180 degrees.)

BACKWARDS
The letters will be created backwards as typed.

VERTICAL
Each letter will be inserted directly under the other. Only **.shx** fonts can be used.
VERTICAL text will not display in the **PREVIEW** box.

OBLIQUE ANGLE
Creates letter with a slant, like italic. An angle of 0 creates a vertical letter. A
positive angle will slant the letter forward. A negative angle will slant the letter
backward.

WIDTH FACTOR
This effect compresses or extends the width of each character.
A value less than 1 compresses each character.
A value greater than 1 extends each character.

8. **PREVIEW** your settings

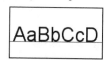

9. Select the **Set Current** button.

10. Select the **Apply** button

HOW TO SELECT A TEXT STYLE

After you have created Text Styles you will want to use them. You must select the Text Style before you use it.

Below are the methods when using Single Line or Multiline text.

Single Line Text

Select the style before selecting Single Line Text command.

1. Select the **Annotate** tab.
2. Using the Text panel, select the style down arrow ▼
3. Select the Text Style

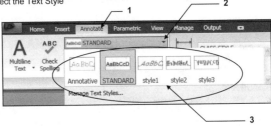

Multiline Text

Select the Text Style within the Text Editor.

1. Find the **Style** panel and scroll through the text styles available using the up and down arrows.

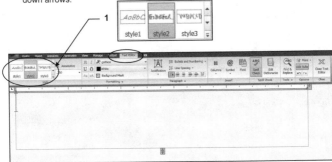

DELETE A TEXT STYLE

1. Select the **TEXT STYLE** command using one of the following:

   Ribbon = Annotate tab / Text panel / ↘

   Keyboard = ST <enter>

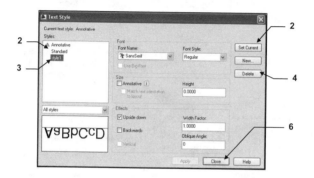

2. First, select a **Text Style** that you <u>do not want to Delete</u> and select **Set Current** button. (You can't Delete a Text Style that is in use.)

3. Select the **Text Style** that you want to Delete.

4. Select the **Delete** button.

5. Warning appears, select **Ok** or **Cancel**

6. Select the **Close** button.

Note: Also refer to the PURGE command. The Purge command will remove any unused text styles, dimension styles, layers and linetypes.

CHANGE EFFECTS OF A TEXT STYLE

1. Select the **TEXT STYLE** command using one of the following:

   Ribbon = Annotate tab / Text panel / ↘

   Keyboard = ST <enter>

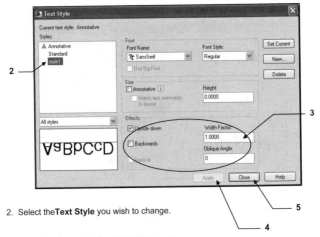

2. Select the **Text Style** you wish to change.

3. Make the changes in the **EFFECTS** boxes.

   Note about Vertical:
   Only **.shx** fonts can be vertical
   Vertical text will not display in the **PREVIEW** box.

4. Select the **Apply** button. (*Apply will stay gray if you did not change a setting.*)

5. Select the **Close** button.

MULTILINE TEXT

MULTILINE TEXT command allows you to easily add a sentence, paragraph or tables. The Mtext editor has most of the text editing features of a word processing program. You can underline, bold, italic, add tabs for indenting, change the font, line spacing, and adjust the length and width of the paragraph.

When using MText you must first define a text boundary box. The text boundary box is defined by entering where you wish to start the text (first corner) and approximately where you want to end the text (opposite corner). It is very similar to drawing a rectangle. The paragraph text is considered one object rather than several individual sentences.

USING MULTILINE TEXT

1. Select the MULTILINE TEXT command using one of the following:

    Ribbon = Annotation tab / Text panel /

    Keyboard = MT <enter>

The command line will lists the current style, text height and annotative setting.

   Mtext Current text style: "STANDARD"   Text height: .250   Annotative: No

The cursor will then appear as crosshairs with the letters "abc" attached. These letters indicate how the text will appear using the current font and text height.

2. Specify first corner: *Place the cursor at the upper left corner of the area where you want to start the new text boundary box and press the left mouse button. (P1)*

3. Specify opposite corner or [Height / Justify / Line Spacing / Rotation / Style / Width / Columns]: *Move the cursor to the right and down (P2) and press left mouse button.*

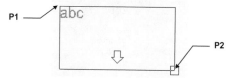

MULTILINE TEXT....continued

The Text Formatting Tool bar will appear.

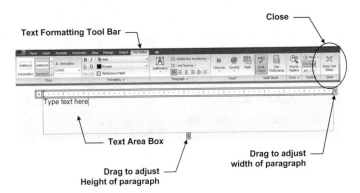

The **Text Formatting toolbar** allows you to select the Text Style, Font, Height etc. You can add features such as bold, italics, underline and color.

The **Text Area box** allows you to enter the text, add tabs, indent, adjust left hand margins and change the width and height of the paragraph.

4. After you have entered the text in the Text Area box, select the **Close Text Editor** tool.

HOW TO CHANGE THE "abc", ON THE CROSSHAIRS, TO OTHER LETTERS.

You can personalize the letters that appear attached to the crosshairs using the **MTJIGSTRING** system variable. (10 characters max) The letters will simulate the appearance of the font and height selected but will disappear after you place the lower right corner (P2).

1. Type **MTJIGSTRING** <enter> on the command line.
2. Type the new letters <enter>.

The letters will be saved to the computer, not the drawing. They will appear anytime you use Mtext and will remain until you change them again.

TABS, INDENTS and SPELLING CHECKER

TABS

Setting and removing Tabs is very easy.

The default setting for tabs is 1". (You may set as many tabs as you need.)
Set or change the stop positions at anytime, using one of the following methods.

Place the cursor on the "Ruler" where you want the tab and left click. A little dark "**L**"
will appear. The tab is set.
If you would like to remove a tab, just click and drag it off the ruler and it will disappear

INDENTS

Sliders on the ruler show indention relative to the left side of the text boundary box.
The top slider indents the first line of the paragraph, and the bottom slider indents the
other lines of the paragraph.

You may change their positions at anytime, using one of the following methods.

Place the cursor on the "Slider" and click and drag it to the new location.

SPELLING CHECKER

If you have Spell check ON you will be alerted as you enter text with a red line under
the misspelled word. Right click on the word and AutoCAD will give you some choices.

You may also check the spelling of your entire drawing using AutoCAD's Spelling
Checker.

1. Select **Annotation tab / Text panel ◢ /**

   The Check Spelling dialog box will appear.

2. Select Start.
   If AutoCAD finds any words misspelled it will suggest a change.
   You may select **Change** or **Ignore**.
   When finished a message will appear stating "**Spelling Check Complete**".

COLUMNS

STATIC COLUMNS

1. Right click in the **Text Box Area** and select **Columns**

2. Select **Column Settings...**

The Column Dialog box appears.

3. Select **Static Columns**

4. Select:
   A. Column Number
   B. Height
   C. Width
   D. Gutter

5. Select the **OK** button

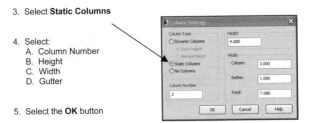

6. The Text Area should appear as shown below with 2 columns divided with a gutter.

7. Start typing in the left hand box. When you fill the left hand box the text will start to spill over into the right hand box.

You may also adjust make changes to width and height using the drag tools.

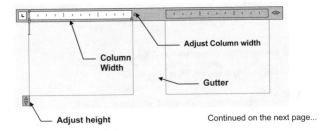

Column Width

Adjust Column width

Gutter

Adjust height

Continued on the next page...

COLUMNS....continued

DYNAMIC COLUMNS

1. Right click in the **Text Box Area** and select **Columns**

2. Select **Column Settings...**

The Column Dialog box appears.

3. Select **Dynamic Columns**

4. Select:
   A. Height
   B. Width
   C. Gutter

5. Select the **OK** button

6. The Text Area will first appear with one column with the width and height you set.

7. When the text fills the first column another column will appear. When the second

column fills another column will appear.
You may also adjust make changes to width and height using the drag tools.

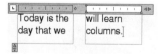

PARAGRAPH and LINE SPACING

PARAGRAPH and LINE SPACING

You may set the tabs, indent and line spacing for individual paragraphs.

1. Right click in the **Text Box Area** and select **Paragraph**.

 The Paragraph dialog box will appear.

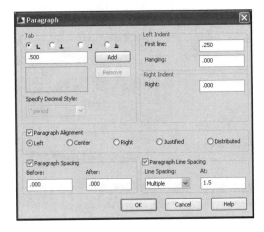

EDITING MULTILINE TEXT

MULTILINE TEXT

Multiline Text is as easy to edit as it is to input originally. You may use any of the text options shown on the Text Editor tab.

1. Double click on the Multiline text you want to edit.

2. Highlight the text, that you want to change, using click and drag.

3. Make the changes then select the **Close Text Editor** tool.

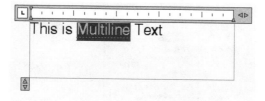

EDITING MULTILINE TEXT....continued

You may edit many Multiline Text features.

Right click in the **Text Box Area.**

The menu shown below will appear.

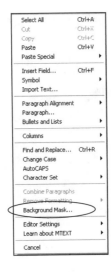

SINGLE LINE TEXT

SINGLE LINE TEXT allows you to draw one or more lines of text. The text is visible as you type. To place the text in the drawing, you may use the default **START POINT** (the lower left corner of the text), or use one of the many styles of justification described on the next page.

USING THE DEFAULT START POINT

1. Select the **SINGLE LINE TEXT** command using one of the following:

    Ribbon = Annotation tab / Text panel / [A]

    Keyboard = DT <enter>

Command: _dtext
Current text style: "STANDARD" Text height: 0.250 Annotative: No
Specify start point of text or [Justify/Style]: *Place the cursor where the text should*
start and left click.
Specify height <0.250>: *type the height of your text <enter>*
Specify rotation angle of text <0>: *type the rotation angle then <enter>*
Enter text: *type the text string; press enter at the end of the sentence*
Enter text: *type the text string; press enter at the end of the sentence*
Enter text: *type the next sentence or press <enter> to stop*

USING JUSTIFICATION

If you need to be very specific, where your text is located, you must use the Justification option. For example if you want your text in the middle of a rectangular box, you would use the justification option "Middle".

The following is an example of Middle justification.

1. Draw a Rectangle 6" wide and 3" high.
2. Draw a Diagonal line from one corner to the diagonal corner.
3. Select the SINGLE LINE TEXT command
    Command: _dtext
    Current text style: "STANDARD" Text height: 0.250
4. Specify start point of text or[Justify/Style]: *type "J"<enter>*
5. Enter an option [Align/Fit/Center/Middle/Right/TL/ TC/TR/ ML/MC/MR/BL/BC/BR]: *type M <enter>*
6. Specify middle point of text: *snap to the midpoint of the*
    diagonal line
7. Specify height <0.250>: *1 <enter>*
8. Specify rotation angle of text <0>: *0 <enter>*
9. Enter text: *type: HHHH <enter>*
10. Enter text: *press <enter> to stop*

SINGLE LINE TEXT....continued

OTHER JUSTIFICATION OPTIONS:

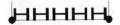

ALIGN
Aligns the line of text between two points specified.
The height is adjusted automatically.

FIT
Fits the text between two points specified.
The height is specified by you and does not change.

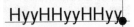

CENTER
This is a tricky one. Center is located at the bottom center of Upper Case letters.

MIDDLE
If only uppercase letters are used **MIDDLE** is located in the middle, horizontally and
vertically. If both uppercase and lowercase letters are used **MIDDLE** is located in the
middle, horizontally and vertically, but considers the lowercase letters as part of the
height.

RIGHT
Bottom right of upper case text.

TL, TC, TR
Top left, Top center and Top right of upper and lower case text

ML, MC, MR
Middle left, Middle center and Middle right of upper case text.
(Notice the difference between "Middle" and "MC"

BL, BC, BR
Bottom left, Bottom center and Bottom right of lower case text.
Notice the different location for **BR** and **RIGHT** shown above.
BR considers the lower case letters with tails as part of the height.

EDITING SINGLE LINE TEXT

SINGLE LINE TEXT

Editing **Single Line Text** is somewhat limited compared to Multiline Text. In the example below you will learn how to edit the text within a Single Line Text sentence.

1. Double click on the <u>Single Line text</u> you want to edit. The text will highlight.

2. Make the changes in place then press <enter> <enter>.

SPECIAL TEXT CHARACTERS

Characters such as the *degree symbol (°)*, *diameter symbol (Ø)* and the
plus / minus symbols (±) are created by typing **%%** and then the appropriate "**code**"
letter.

SYMBOL		CODE
Ø	Diameter	%%C
°	Degree	%%D
±	Plus / Minus	%%P

SINGLE LINE TEXT
When using "Single Line Text", type the code in the sentence. After you enter the
code the symbol will appear.

> **For example:**
> Entering 350**%%D** will create: **350°**. The **"D"** is the **"code"** letter for degree.

MULTILINE TEXT
When using Multiline Text you may enter the code in the sentence or you may select a
symbol using the **Symbol tool** located on the **Insert panel.**

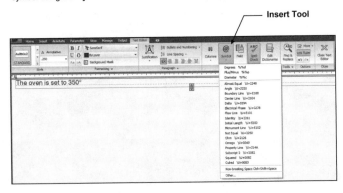

Section 12
UCS and Origin

DISPLAYING THE UCS ICON

The UCS icon is merely a drawing aid. You can control how it is displayed.
It can be visible (on) or invisible (off). It can move with the Origin or stay in the default location. You can even change its appearance.

1. Select the **View tab / Coordinates panel**

2. Select the **Show UCS Icon at Origin ▼**

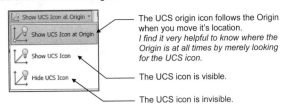

The UCS origin icon follows the Origin when you move it's location.
I find it very helpful to know where the Origin is at all times by merely looking for the UCS icon.

The UCS icon is visible.

The UCS icon is invisible.

3. Select **UCS Icon Properties**

When you select this option the dialog box shown below will appear.
You may change the Style, Size and Color of the icon at any time.
Changing the appearance is personal preference and will not affect the drawing or commands.

4. Select the **OK** button.

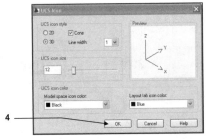

MOVING THE ORIGIN

The **ORIGIN** is where the X, Y, and Z-axes intersect. The Origin's (0,0,0) default
location is in the lower left-hand corner of the drawing area.
But you can move the Origin anywhere on the screen using the UCS command.
(The default location is designated as the "**World**" option or WCS. When it is moved it
is UCS, User Coordinate System.)

You may move the Origin many times while creating a drawing. This is not difficult and
will make it much easier to draw objects in specific locations.

Refer to the examples on the next page.

To MOVE the Origin:

1. Select one of following:

   Ribbon = View tab / Coordinates panel /

   Keyboard = ucs <enter>

   Current ucs name: *WORLD*
2. Specify origin of UCS or [Face/NAmed/OBject/Previous/View/World/X/Y/Z/ZAxis]
   <World>: _o

3. Specify new origin <0,0,0>: *type coordinates or use the cursor to place new
   Origin location.*

To RETURN the Origin to the default "World" location (the lower left corner):

1. Select one of the following:

   Ribbon = View tab / Coordinates panel /

   Keyboard = ucs <enter> w <enter>

MOVING THE ORIGIN....continued

The following are a few examples of moving the Origin.

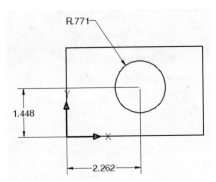

If you move the Origin to the lower left corner of the rectangle it will make if very easy to accurately place the center of the circle.

How to place the Circle accurately:
1. Select the "Move Origin" as shown on the previous page
2. Snap to the lower left corner of the rectangle using object snap Endpoint.
   The UCS icon should now be displayed as shown above.
3. Select the Circle command
4. Enter the coordinates to the center of the circle (2.262, 1.448 <enter>)
5. Enter the radius (.771 <enter)

6. Select UCS World to return the UCS icon to the lower left corner.
   (Refer to previous page)

More examples of moving the Origin

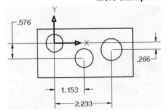
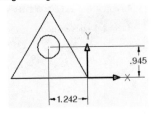

Section 13
Constraints

Parametric Drawing

Parametric drawing is a method of assigning a constraint to an object that is controlled by another object. For example, you could put a constraint on line #1 to always be parallel to line #2. Or you could put a constraint on diameter #1 to always be the same size as diameter #2. If you make a change to #1 AutoCAD automatically makes the change to #2 depending on which constraint has been assigned.

There are two general types of constraints:

    Geometric: Controls the relationship of objects with respect to each other. The Geometric constraints that will be discussed in this workbook are coincident, collinear, concentric, fix, parallel, perpendicular, horizontal, vertical, tangent, symmetrical and equal.

    Example:

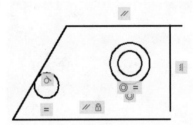

    Dimensional: Dimensional constraints control the size and proportions of objects. They can constrain:
      A. Distances between objects or between points on objects
      B. Angles between objects or between points on objects
      C. Sizes of arcs and circles.

    Example:

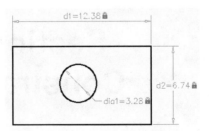

Geometric Constraints

Overview of Geometric Constraints.
Geometric constraints control the relationship of objects with respect to each other.
For example, you could assign the parallel constraint to 2 individual lines to assure that
they would always remain parallel.

The following is a list of Geometric Constraints available and their icons:

You can apply geometric constraints to 2D geometric objects only. Objects cannot be
constrained between model space and paper space.

When you apply a constraint, two things happen:
1. The object that you select adjusts automatically to conform to the specified constraint.
2. A gray constraint icon displays near the constrained object.

Example:
If you apply the **Parallel** constraint to lines A and B you can insure that line B will always
be parallel to line A.

Before

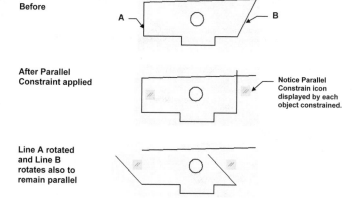

**After Parallel
Constraint applied**

Notice Parallel
Constrain icon
displayed by each
object constrained.

**Line A rotated
and Line B
rotates also to
remain parallel**

Geometric Constraints....continued

How to apply Geometric Constraints

The following is an example of how to apply the Geometric Constraint **Parallel** to 2D objects.
The remaining Geometric Constraints are described on the following pages.

1. Draw the objects.
   Geometric constraints must be applied to <u>existing</u> geometric objects.

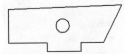

2. Select the **Parametric** tab.

— 2

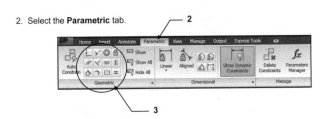

— 3

3. Select the Geometric constraint tool from the Geometric panel.
   (In this example the parallel constraint has been selected.)

4. Select the objects that you want to constrain.
   In most cases the order in which you select two objects is important. Normally the second object you select adjusts to the first object. In the example shown below Line A is selected first and Line B is selected second. As a result Line B adjusts to Line A.

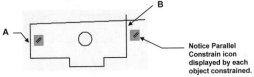

Notice Parallel
Constrain icon
displayed by each
object constrained.

Geometric Constraints....continued

Coincident Constraint
A coincident constraint forces two points to coincide.

1. Draw the objects.

2. Select the Coincident tool from the Geometric panel.

3. Select the first Point.
   (Remember it is important to select the points in the correct order. The first point
   will be the base location for the second point)

   ——— First point

4. Select the second Point.
   (Remember it is important to select the points in the correct order. The second
   point will move to the first point)

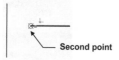

   ——— Second point

The points are now locked together. If you move one object the other object will move
also and the points will remain together.

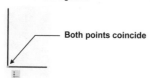

   ——— Both points coincide

Geometric Constraints....continued

Collinear Constraint
A collinear constraint forces two lines to follow the same infinite line.

1. Draw the objects.

2. Select the Collinear tool from the Geometric panel.

3. Select the first line.
   (The first line selected will be the base and the second line will move.)

4. Select the second line.
   (The second line will move in line with the first line selected)

The two lines are now locked in line. If you move one line the other line will move also and they will remain collinear.

Geometric Constraints....continued

Concentric Constraint
A concentric constraint forces selected circles, arcs, or ellipses to maintain the same center point.

1. Draw the objects.

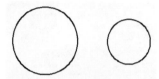

2. Select the Concentric tool from the Geometric panel.

3. Select the first circle.
   (The first circle selected will be the base and the second circle will move.)

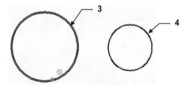

4. Select the second circle.

The second circle moves to have the same center point as the first object.

Geometric Constraints....continued

Fixed Constraint

A Fixed constraint fixes a point or curve to a specified location and orientation relative to the Origin (World Coordinate System, WCS).

1. Draw the object.

2. Select the Fixed tool from the Geometric panel.

3. Select the Fixed location.

The bottom left corner is now fixed to the specified location but the other three corners can move.

 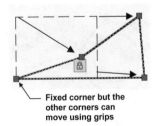

Fixed corner but the other corners can move using grips

Geometric Constraints....continued

Perpendicular Constraint
A Perpendicular constraint forces two lines or polyline segments to maintain a 90 degree angle to each other.

1. Draw the objects.

2. Select the Perpendicular tool from the Geometric panel.

3. Select the First object.
   (The first object will be the base angle and the second will rotate to become perpendicular to the first)

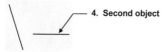

3. First object

4. Select the Second object.

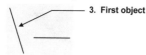

4. Second object

The second object rotated to become perpendicular to the first object.

Geometric Constraints....continued

Horizontal Constraint
The Horizontal constraint forces a line to remain <u>parallel to the **X-axis**</u> of the current UCS.

1. Draw the objects.

2. Select the Horizontal tool from the Geometric panel.

3. Select the <u>object</u> .

select object

The object selected become horizontal to the X-Axis of the current UCS.

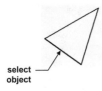

Note: If the object is a Polyline or Polygon use the "2 point" option. The first point will be the pivot point.

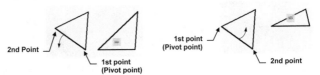

2nd Point

1st point
(Pivot point)

1st point
(Pivot point)

2nd point

Geometric Constraints....continued

Vertical Constraint

The Vertical constraint forces a line to remain <u>parallel to the **Y-axis**</u> of the current UCS.

1. Draw the objects.

2. Select the Vertical tool from the Geometric panel.

3. Select the <u>object</u> .

select —
object

The object selected become <u>vertical to the Y-Axis of the current UCS</u>.

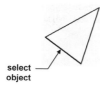

Note: If the object is a Polyline or Polygon use the "2 point" option. The first point will define the pivot point.

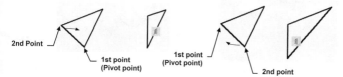

2nd Point —

1st point
(Pivot point)

1st point
(Pivot point)

2nd point

Geometric Constraints....continued

Tangent Constraint
The Tangent constraint forces two objects to maintain <u>a point of tangency</u> to each other.

1. Draw the objects.

2. Select the Tangent tool from the Geometric panel.

3. Select the base object (Large Circle) and then the object (Line) to be tangent.

4. Select the base object (Large Circle) and then the object (Small Circle) to be tangent.

Geometric Constraints....continued

Symmetric Constraint
The Symmetric constraint forces two objects on a object to maintain <u>symmetry about a</u> <u>selected line.</u>

1. Draw the objects.

2. Select the Tangent tool from the Geometric panel.

3. Select the base line (1) then the Line (2) to be symmetrical and then the line (3) to be symmetrical about.

4. Select the base line (1) then the Line (2) to be symmetrical and then the line (3) to be symmetrical about..

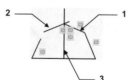

Geometric Constraints....continued

Equal Constraint
The Equal constraint forces two objects to be equal in size. (Properties are not changed)

1. Draw the objects.

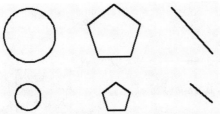

2. Select the Equals tool from the Geometric panel.

3. Select the base object (A) then select the object (B) to equal the selected base object.

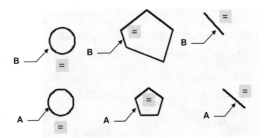

Geometric Constraints....continued

Controlling the display of Geometric constraint icons
You may temporarily Hide the Geometric constraints or you may show individually
selected constraints using the Show and Hide tools.

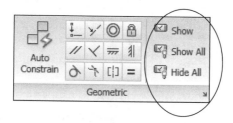

Show All
This tool displays all geometric constraints.
Click on the tool and the constraints appear.

Hide all
This tool hides all geometric constraints.
Click on the tool and the constraints disappear.

Show
After you have selected the Hide All tool to make the constraints disappear, you may
display individually selected geometric constraints.
1. Select the Show tool
2. Select the object
3. <enter>
The geometric constraints for only the selected objects will appear.

Dimensional Constraints

Overview of Dimensional Constraints
Dimensional constraints determine the distances or angles between objects, points on objects, or the size of objects.

Dimensional constraints include both a name and a value.

Dynamic constraints have the following characteristics:
- A. Remain the same size when zooming in or out
- B. Can easily be turned on or off
- C. Display using a fixed dimension style
- D. Provide limited grip capabilities
- E. Do not display on a plot

There are 7 types of dimensional constraints. (They are similar to dimensions)
1. Linear
2. Aligned
3. Horizontal
4. Vertical
5. Angular
6. Radial
7. Diameter

The following is an example of a drawing with dimensional constraints. The following pages will show you how to add dimensional constraints and how to edit them.

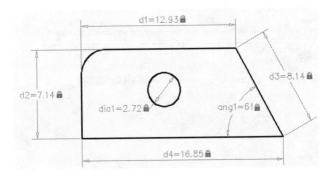

Dimensional Constraints....continued

How to apply Dimensional Constraints

The following is an example of how to apply the dimensional constraint Linear to a 2D object.

1. Draw the objects.

2. Select the Parametric tab.

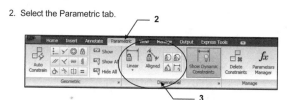

3. Select the Dimensional Constraint tool from the Dimensional panel.
   (In this example the Linear tool has been selected)

4. Apply the "Linear" dimensional constraint as you would place a linear dimension.
   a. Place the first point
   b. Place the second point
   c. Place the dimension line location

5. Enter the desired value or <enter> to accept the displayed value.
   (Notice the constraint is highlighted until you entered the value)

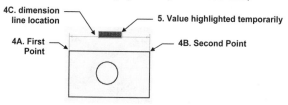

6. The Dimensional constraint is then displayed with Name (d1) and Value (12.38).

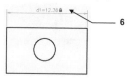

Dimensional Constraints....continued

How to use Dimensional Constraints

The following is an example of how to use dimensional constraint to change the
dimensions of a 2D object.

1. Draw the objects.
   (Size is not important. The size will be changed using the dimensional constraints.)

2. Apply Dimensional constraints
   Linear and Diameter.   — **2**

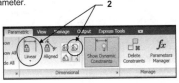

Note: Dimensional constraints are very faint and will not print.

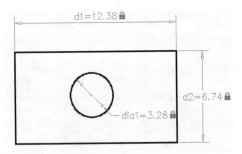

Continued on the next page...

Dimensional Constraints….continued

Now I will adjust the length and width using the dimensional constraints.

1. Double click on the d1 dimensional constraint.

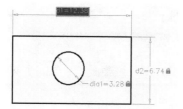

2. Enter the new value for the length then <enter>

Note: The length increased automatically in the direction of the "2nd endpoint". If you want the length to change in the other direction you must apply the "geometric constraint Fixed" to the right hand corners.

3. Double click on the d2 dimensional constraint and enter the new value for the width then <enter>

Dimensional Constraints....continued

Parameter Manager

The Parameter Manager enables you to manage dimensional parameters. You can change the name, assign a numeric value or add a formula as its expression.

Select the Parameter Manager from the Parametric tab / Manage Panel

Name column: Lists all of the dimensional constraints. The order can be changed to ascending or descending by clicking on the up or down arrow.

Expression column: Displays the numeric value or formula for the dimension.

Value column: Displays the current numeric value.

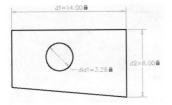

Dimensional Constraints....continued

Parameter Manager Name cell

You may change the name of the dimension to something more meaningful. For example you might change the name to Length and width rather than d1 and d2.

1. Double click in the name cell that you want to change.

2. Type the new name and <enter>

Example:

The names in the Parameter Manager shown below have been changed. Notice the dimensional constraints in the drawing changed also.

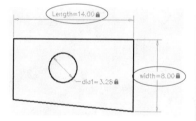

Dimensional Constraints….continued

Parameter Manager Expression cell
You may change the value of a dimensional constraint by clicking on the Expression cell and entering a new value or formula.

1. Open the Parameter Manager. (Refer to previous page)

2. Select the Expression cell to change (dia1 in this example)

3. Enter new value or formula.
   For this example: width / 2
   This means the diameter will always be half the value of d2 (width)

Now whenever the width is changed the diameter will adjust also.

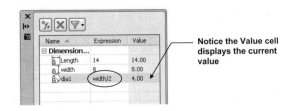

Notice the Value cell displays the current value

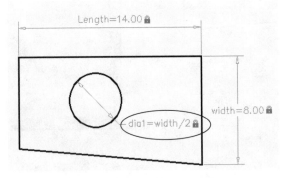

Dimensional Constraints....continued

Adding User-defined parameters
You may create and manage parameters that you define.

1. Open the Parameter Manager

2. Select the "New user parameter" button.
   A "**User Variable**" will appear.

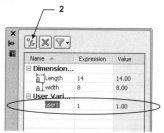

3. Enter a **Name** and an **Expression**.
   The <u>Value cell</u> updates to display the current value.

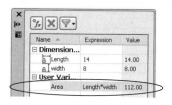

Note: With Imperial Units, the parameter manager interprets a minus or a dash (-) as a unit separator rather than a subtraction operation. To specify subtraction, include at least one space before or after the minus sign. For example: to subtract 9" from 5', enter **5' -9"** rather than **5'-9"**

Dimensional Constraints....continued

Convert Dimensional constraint to an Annotational constraint
Geometric and Dimensional constraints do not plot. If you would like to plot them you must convert them to an Annotational constraint.

1. Select the constraint to convert. (Click on it once)

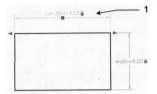

2. Right click and select **properties** from the list.

3. Select the Constraint Form down arrow and select Annotational

4. The Properties palette is populated with additional properties as the constraint is now an annotational constraint.

5. The annotational constraint will not plot.

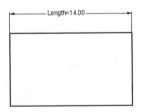

Dimensional Constraints....continued

Control the Display of Dimensional constraints
You may turn off the display of Dimensional constraints using the Show Dynamic Constraints button. Click once to turn OFF. Click again to turn ON.

Delete a Dimensional constraint.
To permanently delete a dimensional constraint select the Delete Constraints button and then select the constraint to delete.

Notes

INDEX